Advances in the science and technology of ocean management

The last two decades have seen unprecedented developments in marine science and technology. Large-scale, international and long-term scientific programmes have emerged for monitoring the state of the ocean environment and the rapid development of the offshore oil industry has provided the technological means for a range of other maritime developments. The purpose of this book is to review key developments in this field.

Three major themes are developed throughout the book: the key importance of technical developments in ocean management; the application of these developments to specific sea uses, ranging from fish farming to the deep sea disposal of industrial waste; and the long-term general issues raised – and to some extent solved – by science and technology.

Hance D. Smith is a senior lecturer in Maritime Studies at the University of Wales College of Cardiff. He is a specialist in teaching and research on ocean management, with particular reference to integrated management approaches. His publications include *The North Sea: Sea Use Management and Planning* and *Oceans and Seas*.

Ocean management and policy series
Edited by H.D. Smith

Advances in the science and technology of ocean management

Edited by
Hance D. Smith

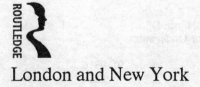

London and New York

First published 1992
by Routledge
11 New Fetter Lane, London EC4P 4EE

Simultaneously published in the USA and Canada
by Routledge
a division of Routledge, Chapman and Hall, Inc.
29 West 35th Street, New York, NY 10001

Typeset in Times by
NWL Editorial Services, Langport, Somerset TA10 9DG

Printed and bound in Great Britain by
Mackays of Chatham PLC, Chatham, Kent

British Library Cataloguing in Publication Data
Advances in the science and technology of ocean management.
   1. Ocean. Use of technology
   I. Smith, Hance D.
   623.87
   ISBN 0–415–03817–0

Library of Congress Cataloging in Publication Data
Advances in the science and technology of ocean
   management / [edited by] Hance D. Smith.
      p. cm.
   "Proceedings of the Challenger Society/Society for
Underwater Technology Conference, UWIST, Cardiff,
12–13 April 1988"—
   Includes bibliographical references and index.
   ISBN 0–415–03817–0
   1. Marine Resources—Congresses.   2. Ocean
engineering—Congresses.   I. Smith, Hance D.
II. Challenger Society.   III Society for Underwater
Technology.
GC1015.2.A36 1991   91–9516
333.91'6515 – dc20   CIP

Contents

Illustrations

FIGURES

PLATE

Tables

Contributors

Martin V. Angel
Deacon Laboratory, Institute of Oceanographic Sciences, Godalming

C.S. Barr
Hydrographic Department, Ministry of Defence (Navy), Taunton

K.R. Deeming
Metocean Consultancy Ltd, Haslemere

R. Earll
Marine Conservation Society, Ross-on-Wye

John Gibson
Centre for Marine Law and Policy, Cardiff Law School, University of Wales, College of Cardiff

M.J.S. Gillespie
Marine Farming Unit, Sea Fish Industry Authority, Ardtoe

S. Gubbay
Marine Conservation Society, Ross-on-Wye

J.E. Halliday
Department of Geography, University of Exeter

Geoffrey L. Haskins
Metocean Consultancy Ltd, Haslemere

A.G. Hopper
Sea Fish Industry Authority, Hull

D.A. Huntley
Institute of Marine Studies, Polytechnic South West, Plymouth

John M. Huthnance
Proudman Oceanographic Laboratory, Birkenhead

R. McGarvey
Department of Biology, Dalhousie University, Halifax, Nova Scotia

J.M. McGlade
University of Cambridge and Max Planck Institute, Cologne

F.G. Parrish
Crown Estate Commissioners, London

Adrian F. Richards
Adrian F. Richards Company, Aalsmeer, and Delft University of Technology

A. Ross
Marine Conservation Society, Ross-on-Wye

Hance D. Smith
Department of Maritime Studies, University of Wales College of Cardiff

P. Wainwright
Hydrographic Department, Ministry of Defence (Navy), Taunton

John Yates
Programme of Policy Research in Engineering, Science and Technology (PREST), University of Manchester

Preface

The work of this international conference on sea use management arose out of the perceived need for co-operation in the integration of science and technology applied to the new, rapidly developing fields of ocean development and management. This led to the Challenger Society and Society for Underwater Technology getting together to examine the nature of such co-operation and integration.

The initial approach was to consider ocean management in terms of a series of technical and general management functions applied to the major groups of sea uses. The technical management functions include scientific fields (monitoring, surveillance, information technology); technology fields (technology assessment and project development) and environmental fields (resource, risk and environmental impact assessment); together with general management considerations dealing with conflicting uses, integration of technical management functions and overall policy considerations.

The papers submitted were readily subdivided into seven working sessions, dealing respectively with the sea bed (resource assessment, development and management), fisheries and fish farming, the ocean waters (modelling and information technology), multiple uses and information technology, integrated management approaches, and coastal and sea use management.

In reviewing the work of the conference it became apparent that three major thrusts are evident. The first is concerned with technical developments, both in field studies and in information technology. The emphasis here is very much on the sea bed itself as far as field studies are concerned, which contrasts with the information technology aspects focused on the uses and environment of the sea itself. In the first case the forefront was represented by the continuing development of remote sensing techniques and special-purpose charting technology (not published fully here); in the second the key themes of digital information handling, data base management and modelling are covered.

The second major thrust of the conference was concerned with the management of individual uses. This is especially important, because the history of sea use management to date is largely written in terms of individual use techniques and organization. The marine environment and its living resources are considered first. The fish farming paper looks forward to the fastest-developing area in industrial countries' marine living resource production. The following two papers are concerned with forward-looking areas of use of the marine environment as a whole. The pressure for marine waste disposal, especially of sewage sludge, remains very great, and the novel idea of using the deep ocean is likely to become increasingly attractive to urban industrial regions. Marine conservation, in the ecosystem sense, is the newest of the sea use groups, and perhaps fittingly begins at the coast, where the environmental impact pressures are usually greatest, and the marine environment is most accessible to the human population for other uses.

The development of mineral resources is of course one of the leading areas of sea use development, and perhaps the major force in encouraging integrated management approaches, not least through the impact of the oil and gas industry. Rather than consider oil and gas, however, the conference focused first on the management revolution which the long-established sand and gravel dredging industry in the North Sea is undergoing, and which is instructive in the development of hydrocarbons and other marine mineral resource development and management. This is followed by a look at the polymetallic sulphide deposits of the deep ocean, which may well be more attractive than the fabled 'manganese nodules' for initial mineral exploitation in this environment.

The final major thrust belongs to the complex field of general management, where two main themes are immediately apparent. The first concerns strategic, long-term decision-making, which is of key importance in the essentially large-scale enterprises of mineral resource exploitation, fisheries research and development, and marine science programmes, for example. This is a practical section, which reviews long-term strategy in marine science, technology and the technology strategy approach, moving on to consider practical experience in the Canadian coastal sediment study, and the crucial area of management of data-gathering programmes.

The second theme is more conceptual in approach. The regional context is basic to most management decision-making. The first two papers focus primarily upon the coastal zone, which is arguably the first great challenge for the application of science and technology in ocean management, owing to the pressures of conflicting uses and environmental impact, together with the need for integration with land management systems. Both conceptual aspects and a case study are considered, and a similar approach is adopted in the following two papers, which are concerned with the offshore areas beyond the coastal zone. The Dutch research programme (not published here), for the management of that country's exclusive zone is especially significant, as the first of its kind, while the modelling paper tackles concepts and practice which will lie behind most management approaches, not just in the fisheries as considered here.

As is appropriate in a newly developing field, much of the discussion proceeds with reference to examples and experience. This is the present reality of ocean management, although the long-term aim is moving inevitably towards achieving not only effective integration in the application of science and technology but also basic management objectives, and both general and technical management approaches which are shared widely across the spectrum of sea uses. The conference, in identifying key themes, was a useful first step.

H.D.S.

Acknowledgements

Thanks are due to Martin Angel of the Challenger Society for Marine Science, and to Gordon Senior and David Wardle of the Society for Underwater Technology, in their role of providing initiative and organization for the conference; and to Dr Tracy Ratcliffe and Miss Chantal Nicholls of the Department of Maritime Studies at the University of Wales College of Cardiff for acting as rapporteurs during the proceedings. The help of the Department of Maritime Studies and of the staff of the Aberconway Building, both of the University of Wales College of Cardiff, in hosting the conference is also gratefully acknowledged.

Part I
Technical developments

1 The integration of digital navigation information into multiple-use data bases

Geoffrey L. Haskins

The title may be construed as implying a possibility or need to integrate all the information required to navigate safely, i.e. an electronic chart display and information system (ECDIS) data base, into other multi-use data bases. While there may be a requirement for some or all those data to be incorporated for particular applications, it is also clearly possible for them to be a part of some vast all-embracing marine data base covering all possible applications. However, the chapter has been written with the purpose of bringing to the attention of marine data base compilers, designers and users the importance of the techniques used for determining the position of the information pertinent to their own interests; there is really no necessity for good-quality seismic, oceanographic, biological or hydrographic data to be corrupted by doubt about the quality of the positioning. All data bases should include a reference or statement as to the derivation of the positions quoted.

RECENT DEVELOPMENTS

It was in May 1987 that the International Hydrographic Organization (IHO), at its thirteenth quinquennial conference in Monaco, approved the final report of the IHO Committee on the Exchange of Digital Data (CEDD). The report included the CEDD Format for the Exchange of Digital Data, now to become the IHO format, approved internationally for passing nautical charting information in digital form between member states' hydrographic offices.

For five years the CEDD, under the chairmanship of the United States National Ocean Survey, and drawing heavily upon the expertise of the US Defense Mapping Agency (DMA) and Canadian Hydrographic Service (CHS), had worked steadily towards its deadline of May 1987. Several meetings of the member states' representatives took place in Europe, and test tapes were tried out in hydrographic offices all over the world, the resulting comments being incorporated into the end result. The format selected is basically quite simple, making use of point, linear, area and textual data, together with feature codes and their associated attributes based upon, but considerably expanding, the existing IHO Standard Lists of Symbols and Abbreviations.

Over the same period that this development took place, two other data problems were being investigated: the advent of the so-called 'electronic chart' and the possibility of compiling a sixth edition of the General Bathymetric Chart of the Oceans (GEBCO, produced jointly by the IHO and the International Oceanographic Commission (IOC)). The impact of the former was something of a bombshell in the hydrographic community, which had just (in 1985) finalized the procedures and formats for producing international versions of conventional charts on medium and large scales. The latter study was a follow-on from the successful completion of the fifth edition of GEBCO, currently available in conventional form as bathymetric maps, and was concerned with the possibility of providing future users with their information in digital form on tape or disc.

Already in existence, and located in Canada in the Canadian Hydrographic Service, the IHO Data Bank of Tidal Constituents was regularly used. The stations for which the constants are held are identified by their geographical positions in latitude and longitude, qualified by the number of periods of observations at that position if there are more than one.

Meanwhile, beyond the immediate boundaries of the IHO's direct concern, various organizations were producing, developing or improving their own geographical digital data bases and marine information systems, the identity of which was related to, or expressed as, latitudes and longitudes. Indeed, some of these data bases are concerned with cartography (nautical, marine or otherwise, i.e. the presentation of information on a map) in national and other international bodies. Others are principally designed to provide a storage facility or as input to data manipulation or textual presentation.

Despite all this past and present digital data base activity some fundamental items of information are very frequently bypassed or not even recognized as pertinent by the designers of data bases: these very important, but often missing, factors are the measurements that established, in the first place, the geographical positions being used as identifiers or locators of other information. It is therefore of interest to examine the various phases that marine data pass through to determine when and where some of the data vanish.

THE PHASES

Data acquisition

Hydrographic, oceanographic and seismic data acquired at sea are normally identified by a line and station/shot-point/fix or event number. The 'fix' is determined by the positioning system used, the elements of which are measured and recorded, together with an 'initial' computed position. In the case of seismic surveying for instance, this navigation data may be recorded separately as a navigation data sub-set. A hydrographic survey, on the other hand, might integrate the positioning with water depth, tidal height, ship's head, distance to offset positions, and any other measured data. It is of paramount importance that the positioning data be subject to a careful on and off-line checking procedure (quality control), and therefore the observed data should be preserved, being termed 'raw data'. The basis of all subsequent data sets, whether they are derived directly or indirectly from the original data, must obviously stem from these raw data.

Post-processed raw data

Raw data acquired on-line in the field are subject to post-mission checking and editing. The data discarded at this stage very often consist of the measured parameters for positioning (being replaced by the computed and checked geographical position) and are also a function of their success in passing the checking procedures or proximity to other data (over-writing). Yesterday's concepts termed this stage the 'fair chart' or 'smooth sheet' – a graphical representation of the results of a survey. Today the data set, together with its quality control (QC) flags, headers describing the data, and trailers, becomes a fully checked basic data set for a specific area and purpose, ready to be incorporated into a principal data base. In the case of seismic data, it is the 'shot-point tape'. For any third party able to read the format of the data, it may be used for exchanging data. However, in some formats it may also be too bulky for further practical applications, and require further selection or weeding out. Conventional two-dimensional seismic position data will be used to plot final shot-point maps and/or be merged into other specialized data processes at this stage. In many cases, unfortunately, the navigation data will be discarded once a position has been calculated, whether or not checks have been successfully applied.

Cartographic selection

The compiler of a map or chart uses a multitude of different sources: data sets from surveys made at different times, already published cartographic or general information, supporting documentation, etc. He probably still compiles his map conventionally by hand, but is today (1988) making increasing use of interactive computer graphics. In both cases the data selected will be only a very small proportion of the original data actually contained in the processed data set. For instance, water depths are largely replaced by contours. The resultant digital end product is the cartographic data set for a specific area and pertinent to a specific scale (the two primary factors which control the data selection). In the case of seismic data, they will not enter this stage at all.

Digital map production

While it is, in theory, quite feasible to produce a high-quality final map or chart by digital means from the original survey data (raw data set) to the final sheets from which the printing plates are made, this is seldom actually achieved. What in fact happens is that a hand-compiled chart is digitized and the resultant data set used for plotting the final sheets on a flat-bed automated plotter. Most so-called digital cartographic data consists of data digitized either from compilation sheets or previously published maps and charts (with all the obvious drawbacks as regards accuracy that are entailed).

An organization's collection of digitized cartographic data forms its data bank. Under the circumstances outlined above it has the following deficiencies:

1 There is unlikely to be any reference to the quality of the positioning of the original data, such as the system used, the calibration results, the deviations found, the accuracies achieved.

2 The original depth data: measurements, methods used, calibrations, accuracies may, equally, not be readily available. This is probably true of all measured data.

3 Data digitized for or at a particular scale may become bundled together with disparate data unless each set is carefully flagged and described. There is a danger that data may be used for making presentations at a scale for which they were not intended.

4 There may be a duplication of data in two adjacent data sets which have disparate values or descriptions. Inconsistencies then arise and may be difficult to eradicate, even by digital means.

What is very clear is that expensively acquired positioning and depth data in the form of original survey measurements drop out of sight unless exceptional efforts are made to preserve them. Experience shows that they are often required for reprocessing using different datums, co-ordinate systems, or values for the velocity of the energy in the medium used. The seismic surveying world learned this lesson, and original navigation data are now usually preserved, either by the client operating company or, for a finite period, by the contractor responsible for the survey.

SAVING THE SURVEY MEASUREMENTS

Marine data generally consists of the following:

Hydrographic surveying

Point and linear: fix number, position measurements, computed position, water depth, tidal height, contours and topography.

Textual: descriptive information (lights, buoys, beacons, navaids, etc.), administrative information, quality information.

Oceanography

Point and linear: station number (or fix number in the case of, e.g. bathymetry), computed position, contours, sensor measurements/ depth, sea bed geology.

Textual: quality information.

Seismic

Point and linear: line number, shot-point number, time, position measurements, computed position, water depth, offsets, ship's head, pitch and roll (to correct antenna position).

Textual: quality information, base station co-ordinates, geodetic data, projections and datum, method of computation, method of interval control (time or distance), antenna height and offset measurements, calibration data, filtering details. (It should be noted that the seismic data are merged with the positioning later in the processing stage on the basis of line number, shot-point number and time. Often this process itself can be a source of error.)

Cartographic data sets

Point and linear: position, features and associated attributes, borders and scales.

Area: Contours, vegetation or sea bed characteristics, etc., areal limits containing sets of point and linear data, e.g. sandbanks.

Textual: titles, source diagrams, annotations, place names, notes.

The common features in all these data sets are: event identification or feature code, geographical position and water depth. The question arises: 'Is there any room for all the supporting data that we would like to preserve?' The answer must be 'No, not unless the data set becomes too unwieldy for ready access.' It is therefore deemed to be impracticable for one data set to contain all the data that are desirable, and an alternative approach must be sought on the principle of relational data bases and sub-sets.

THE MEASUREMENT DATA SET

It is the offshore oil and gas industry that has led the way in preserving original positioning data measurements. In the first place the prime need was the 'shot-point (SP) tapes' – virtually a shot-point map in digital form, with line number, SP number, position, and grid/graticule with borders. These were used for permanent storage and exchanging or selling seismic data. In due course, after many thousands of kilometres of marine seismic data had proved to be of no use owing to the uncertainty of their positioning, the preservation of the original measurements *in perpetuo* became necessary.

The United Kingdom Offshore Operators' Association (UKOOA), through the Exploration Committee's Position Fixing Group (now renamed as a committee in its own right), developed formats during the 1970s for the explorers' differing needs. The 'exchange tape' format was preserved, but a format was agreed for the storage of measured positioning data: the UKOOA Raw Data Format. This serves several purposes:

1 Raw data can be cross-checked for errors against a separately recorded set of quality control navigation data.
2 It can be processed in-house and compared (digitally) with the seismic contractor's smoothed and filtered shot-point map data.
3 It can be reprocessed using updated values for base stations, velocity of propagation, offsets, etc., should it become necessary.
4 All or part of the data can be incorporated into a central data bank for future reference. For instance, while all the quality control

information (periodical calibrations, off-line and SP interval statistics, lane checks, cross-fixing, monitor readings, standard deviations, etc.) are retained, perhaps only half or less of the actual computed shot-point data are loaded.

There is no reason whatever, if there is sufficient incentive, for such a data set to be universally applicable to any of the various marine surveys, whether their primary objective be engineering measurements, hydrography, oceanography or seismic profiling. The main point is that it would be separate from, but referred to by, the other data sets. The incentive in the oil industry for UKOOA formats A and B came from the hard school of experience: drilling a well on the basis of wrong information is not only costly, it also leads to senior employees losing well remunerated jobs. In the case of hydrographic surveys, wrongly positioned hazards to navigation can lead to loss of life and property, and positioning is regarded as of primary importance. Perhaps the incentives in other marine data-gathering fields are not so strong.

A UNIVERSAL MEASURED DATA BASE SUB-SET

Having established that there are common features in most marine data bases involving position, and water depth, i.e. measured data of hydrographic or bathymetric importance, there may be a possibility of accumulating this basic measurement information, stored in a common format, in national or even international centres. The World Data Centre, Boulder, Colorado, is already active in this field with regard to processed data (largely digitized from existing maps and plotting sheets). It is also proven technology that such basic data can be successfully banked. With regard to depth data the problem of storing bathymetry measured by swathe sounding systems rather than single sounding profiles is akin to that of the geophysicist coping with three-dimensional seismic data – they may be stored in 'bins' of information. (But where are the 'bins'?)

The advantage of such a system lies in improved facilities for the end users of data. For instance, the ensuing sequence can be imagined:

1 A data user acquires a set of, say, sea bed biological sampling data for determining environmental changes.
2 Data are compared with those acquired five years previously in the same area.
3 Significant changes are noted – but they may be due to critical position-dependent factors. How accurate were the surveys?
4 The surveys refer to the relevant basic positioning measurement data

set, available on call via a computer terminal from a central data bank. It is therefore a simple matter to check.

5 The check reveals that the original survey was made using a low-accuracy positioning system (all details are available on call) and that the latest survey was to a much tighter standard of accuracy.

This scenario may seem a little far-fetched – but in terms of today's technology it is entirely feasible. Centralized data banks are already in existence for all marine data. Local banks of data for positioning and depth measurements are already available in hydrographic offices, oil exploration companies and survey companies. The time may be ripe to weld the two together.

There is another important practical advantage to be gained by actioning this proposal. All over the world vast areas of the continental shelves have been surveyed for oil and gas reserves during the last thirty-five years. The maps of intergovernmental bodies which publish statistics concerning the apparent dearth of bathymetric data on and off the continental shelves show only a very small proportion of the depth information acquired by the seismic surveys, and this for two reasons: the first is that much of the data, if, as is unusual, they are actually rendered to the mapping or charting authorities from these sources, is of such poor and unrecoverable quality that it is useless. The second is a lack of recognition by either party (host coastal state or operating company) that this by-product has any useful role.

Had there been an internationally recognized need to preserve positioning data (and the soundings), the status of bathymetry in exclusive economic zones would not be quite so bad as it is now. The data acquirers would have been 'encouraged' to reach an acceptably high standard of data capture, and their valuable records would not have been lost.

THE FUTURE

Continuous twenty-four-hour satellite positioning, either government or private sector-sponsored, will ensure that good positioning data will become readily accessible to all maritime users. The doubtful positions of the past will still be on the record and cannot be remedied. Swathe acoustic data (multi-beam sounders, long and short-range sonars) will lead to the 'binning' of bathymetric data in pixels of a size that can be merged with satellite-sensed data. Data communications technology will enable central data banks to send selected thematic information to any destination in a brief spurt of radio energy to any user equipped with a suitable terminal. The terminals themselves will decrease in cost until they are readily available to all. Measured data, unsullied by human intervention, will pass through intelligent systems designed to select, filter and quality-control the data up to and including the final stage via satellite links from ship to home base.

This is what the pundits say – but it is to be sincerely hoped that all the pitfalls associated with getting a remotely sensed survey measurement correct *first time around*, and the need to ensure that the original data are not lost to posterity (just in case it wasn't) will be given proper recognition.

The advent of ECDIS has given the international hydrographic community an edge over many of the other data gatherers, storers and presenters in that systems are already operational, albeit without official status of any kind.

World-wide data bases have been compiled by commercial interests who digitize nationally published charts in a wholesale, indiscriminate and, in most cases, unauthorized manner. The user, in this case the mariner, has his navigational position (from satellite or radio determination) combined with the true motion radar picture overlaid on a VDU presentation of a digital charting data base. Supporting data – sailing directions, light lists, radio signals, racon beacon characteristics, reliability diagrams, etc. – can all be recalled from his storage medium (cartridge bubble memory, compact disc, floppy disk) on demand and displayed on a second or split VDU. These existing systems have been developed by ingenious computer professionals with little or no knowledge of nautical cartographic practice or the real needs of the users, no consideration of ultimate responsibility for the data, updating the data, flagging unsuitable data or funding the data acquisition. Above all, they barely recognize that the data they are presenting represent only a tithe of the data that may actually be available. Even so, if all the published data are incorporated into the working data set, the navigator will have no less information than he has available to him in a conventional ship's chartroom; it is simply accessible in a different way.

CONCLUSION

In view of the exponential increase in computing power, communications technology and international technical co-operation, the development of multi-use data bases in respect of marine data in the future should meet the following user needs:

1 Working vector data sets for individual uses should be capable of

accommodating other data from many other sources, and should contain references to those sources.

2 Original field-recorded data, in a checked and verified form, but also capable of reconstitution or reprocessing in the event of data loss elsewhere in the chain, should be stored safely.

3 An internationally agreed positioning data format for original positioning measurements, water depths, fix information, time, date and geodetic data should be available, and perhaps the option of central storage facilities can be investigated. The oil industry's existing format should be considered as an example in terms of current technology.

4 Vector or raster cartographic data sub-sets should include all necessary information to enable an audit trail back to the origin of all the basic information contained therein.

These four conclusions stem from someone who has been an acquirer, processor, user and disseminator of marine cartographic data for many years, but who regards the advent of digital multi-user data bases containing point, linear, areal and pixel data from many differing sources, and for equally many different uses, with misgivings – particularly if he is dependent upon the end product for his safety or livelihood.

2 Data base management in the production of climatic oceanography

P. Wainwright and C.S. Barr

Knowledge of climatic oceanography plays an important part in Royal Navy operations and the Hydrographic Department at Taunton forms part of the overall Ministry of Defence effort to measure, understand and exploit the oceanographic environment. Parameters of prime interest are temperature, salinity and sound velocity, and data bases which indicate the spatial and temporal variation of these through the water column are used in the analysis of ocean climatology.

This chapter outlines the computer facilities available for managing these data banks, and lists the many data sources used, together with the methods of archiving. Some aspects of the way in which the data are used are described, although a detailed description of the final use of the data products is outside the scope of this paper. Plans for improvement of both the services and products are also described.

THE DATA

Computer hardware

The large volumes of oceanographic data involved can be handled only by using a computer. A Data General MV 8000 processor with four megabytes of memory is used for physical oceanography applications. The current configuration of this machine is: four 592 Mbyte fixed discs, one 354 Mbyte fixed disc, two 277 Mbyte exchangeable disc drives, three 1600/6250 b.p.i. magnetic tape drives, one 600 l.p.m. line printer, one dot-matrix printer, eight D460T VDU terminals and a 1635R A0 size Benson drum plotter. This system replaced an elderly ICL 1900 series computer in mid-1986 and has both greatly improved the speed at which data can be processed and provided greater flexibility.

Software

Manipulation of the oceanographic data bank also needs extensive software facilities. A complex system of in-house software is needed to control the processing of oceanographic data. Currently some seventy programs are employed, with a further twenty or so being developed to come into production use in the next two years. Of these programs about two-thirds are concerned with the archiving and manipulation of data, with the remainder dedicated to data analysis and information presentation. In addition extensive use is made of Data General's software packages, including the efficient SORT/MERGE utility for data manipulation and the PRESENT/TRENDVIEW information presentation facility.

At present data listings and graphical plots are used for the examination of oceanographic data held. The determination of representative climatic oceanography involves much manual examination and editing of automatically plotted temperature and salinity profiles. Sparseness of data in many areas prevents fully automatic methods being employed and consequently an experienced oceanographer must be available to inspect the proposed solutions and edit where necessary.

Recent advances in ocean technology have resulted in the proliferation of new types of instruments for taking oceanographic measurements, and this, combined with a diversification of the types of magnetic media employed in recording their readings, has meant that a wide variety of media readers are needed to examine the data.

Data banks

A data base is a collection of data the structure of which is independent of application programs or hardware. The term is commonly used today to refer to the modern relational, hierarchical and network data bases and their associated data base management systems (DBMS). Plans to introduce the first of these are in hand and will be described below. The term 'data bank' is used here to describe the data files in order to avoid any confusion.

The data are currently stored as several sorted sequential disc files. Some of these are on fixed discs and some on exchangeable discs, depending on their frequency of access. The following data are data-banked at present: bathythermograph observations (depth/temperature); serial data, (a) water (Nansen) bottle data (depth/temperature/salinity) and (b) oceanographic probe data (depth/temperature/salinity/sound velocity).

Bathythermograph observations

The Hydrographic Department operates the UK national data bank for BT observations. The largest single file held is the bathythermograph data bank, which now stands at 1.25 million observations. The data are arranged in a variable-length record format to optimize the use of available storage capacity. This format has been compressed as much as possible to avoid empty fields in records but the file is still 276 Mbytes in size.

The data bank is world-wide, but the North Atlantic is the primary area, with the South Atlantic and Indian Ocean next in importance. Data are sorted into ascending Marsden (10-degree square) order with month/day as the secondary key. The year of the observation is a sorting key of minor importance, owing to the usual requirement to access the data as complete data sets for a time period irrespective of the year.

A large proportion (40 per cent) of the data are measurements by mechanical instruments (MBTs), commencing during the 1940s. Data were recorded on smoked glass slides contained within the instrument. When the instrument was winched back on board ship the slide could be replaced by a new one. The accuracy of the measurements was reasonable (0.2°C) but only less than 300 m of the water column could be measured.

By the mid-1960s these instruments were becoming obsolescent, having been replaced by the expendable bathythermograph (XBT). At first these obtained readings of temperature to 460 m but later versions were developed which dropped to 760 m and then 1830 m. Variations of these probes are designed for launching from aircraft. Expendable bathythermograph probes send a signal from their temperature sensor up a thin piece of copper wire to a shipborne recorder. Drop rate equations are used to calculate depth. One of the main advantages of BTs is that they are designed for use by a ship under way at speeds of up to 30 knots.

Serial data

This term is used for both water (Nansen) bottle data and data recorded by oceanographic probes. Observations by the former commenced in the closing years of the last century but by the 1960s probes had been developed which measured conductivity, temperature and pressure. Some of these were also fitted with sound velocity sensors, although sound velocity is easily computed from the other three parameters and latitude, using internationally agreed equations. Some probes converted conductivity to salinity and pressure to depth internally but it is now more usual to calculate these in-office before archiving the data.

The main difference between Nansen bottle and probe data is that probes give a continuous profile of the water column, from the surface to their maximum operating depth, whereas Nansen bottles are attached to cables and provide spot values at pre-selected depths. Generally these approximate to standard depth levels, of which thirty-four define the water column from the surface to 9000 m, but sea state, currents, etc., at the ocean station can cause variations. Missing bottles often cause a problem when attempting to construct a complete profile from this type of data.

Two serial data banks are maintained at Taunton. The high-quality file, so named because it contains the probe data at one-metre depth level observations, may have up to 4949 m depth levels. Should the probe have been lowered to depths greater than 4949 m the archiving programs identify the occurrence and the data are then stored at 2 m increments after 3000 m. This file is small at present but is gradually increasing as more probe observations are taken. Geographically it is much more limited in scope than the other files, as it is heavily biased towards areas of the world where recent oceanographic research and surveys have been conducted.

The larger serial data bank is a combination of water bottle and probe data. In order to incorporate the latter a reduction of the number of depth levels takes place by compressing the profiles to a maximum of ninety-nine inflection points. Details of this procedure are provided below in the section on data archiving. The coverage of this data bank is similar to that of the bathythermograph data and the same sorting order is used. However, the data bank is divided into a number of separate files, each relating to a discrete ocean area. This has been done because of the size of the complete data set. At 0.5 million observations it is smaller than the BT data bank but, although variable-length records are used, many unused fields exist and it is not possible to compress the records further and retain readability with the COBOL programming language. In total these files have a combined size of 552 Mbytes.

With extensive files as described above, inventory and data distribution programs play an important role in the management of the data base. Line-printer inventories and graphical plots are used to inspect existing data holdings and examine incoming exchange tapes. The inventories tabulate counts of observations against any two key fields with paper throw when a third and primary key field changes value. The plots either display point symbols for individual observations or represent degree squares by different symbols, depending on an allocation of observation counts to predetermined bands, e.g. 1–5, 5–10, 10–50, over 50. Figures 2.1–2 show a sample inventory and distribution plot respectively.

```
                        MARS/DEG/MON INVENTORY
                          COMPILED 05/06/87

MARS 370

MON    01  02  30  04  05  06  07  08  09  10  11  12  TOTAL

      JAN FEB MAR APR MAY JUN JUL AUG SEP OCT NOV DEC
DEG
 0     0   1   0   0   0   0   0   0   0   0   2   0    3
 1     0   3   1   0   1   1   0   0   0   4   2   4   16
 2     0   1   0   2   1   5   0   0   0   6   4   4   23
 3     0   0   0   2   1   4   0   0   0   4   3   8   22
10     0   0   1   0   0   1   0   0   0   0   1   2    5
11     0   0   1   0   0   1   0   0   0   3   5   5   15
12     0   2   4   4   5   3   2   0   0   6  11   4   41
13     4   0   1   1   5   4   0   0   4   2   2  10   33
20     0   0   4   0   0   1   2   0   1   0   4   2   14
21     0   1   4   0   2   2   2   0   2   7   7   4   31
22    20  27  20  28  21  18  25  20  31  29  37  28  304
23    20  15  18  20  18  15  20  12  22  21  14  24  219
30     0   0   1   0   1   0   0   0   0   2   6   2   12
31     0   3   3   0   0   0   6   0   0   4  19   4   39
32     1   0   3   1   3   0   7   0   4   8  16   9   52
40     0   0   2   0   0   0   2   0   0   5   5   0   14
41     1   0   8   0   2   0   9   0   0   6  20   2   48
42     0   0   6   0   2   0   6   0   0   3   9   3   29
50     0   0  12   1   0   0   7   0   0   4  12   0   36
51     0   0  19   3   1   0  16  11   0   7  21   5   83
52     2   0   1   1   0   0   2   1   0   0   1   1    9
60     0   0  11   0   2   0   7   0   0   3  10   0   33
61    11   0  19   6   6   0  13   0   0   3  18   0   76
70     2   0   9   0   3   0   6   0   0   2   9   0   31
71     9  10  22  14  12   0  10   0   4   6   6  14  107
80     1   0   6   0   2   0   0   0   0   0   0   0    9
81     5   8   4   8  12   0   1   0   0   1   0   4   43
82     2   0   0   3   2   0   0   0   0   1   0   0    8
90     2   0   1   0   1   0   1   0   0   5   0   0   10
91     3   5   4   5   2   0   1   0   0   5   0   0   25
92     5  12   9   8   9   0   2   0   0   5   0   3   53
Totals 88  88 194 107 114  55 147  44  68 152 244 142 1443
```

Figure 2.1 A sample data inventory
Source: Hydrographic Office

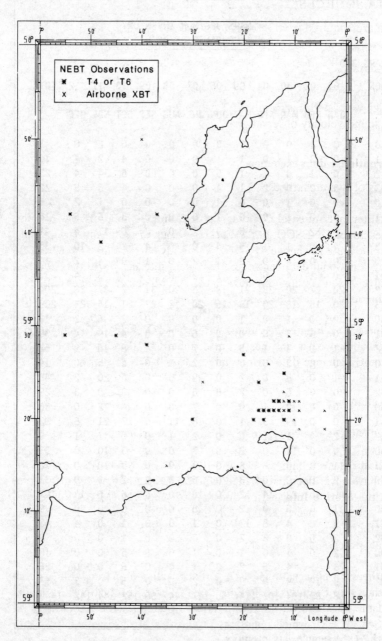

Figure 2.2   An example of a data distribution plot
Source: Hydrographic Office

DATA SOURCES

Data are acquired from three main sources:

Raw data

Raw (unprocessed) data are obtained from the Royal Navy and certain foreign vessels. This is mainly XBT data but may also include STDV/CTD and sound velocity observations.

International data exchange

This is of two main types:

The Intergovernmental Oceanographic Commission (IOC) w h i c h comes under UNESCO, has a Working Group on 'International Oceanographic Data Exchange' (IODE). This body encourages each nation to set up a national data centre and feed data into the world data centres (Centre A is in Washington and Centre B in Moscow). Many useful data can be obtained from this source.

Reciprocal exchange agreements within NATO This is usually done on an annual basis and has been facilitated by the development of a common exchange data format, largely from a UK initiative.

Miscellaneous sources

Other MOD authorities, e.g. Admiralty research establishments.

Civilian research bodies, e.g. the IOS, MAFF. Many very useful UK data have never been data-banked and MOD recently placed a contract with the Marine Information and Advisory Service (MIAS) to collate and process over 21000 CTD/STD observations. Expendable bathythermograph observations from UK civilian research institutes are usually processed as raw data.

Weathership observations. We process all data from the ocean weather ship *Cumulus* at Station 'Lima'. These observations are made available to others via the International Council for the Exploration of the Sea (ICES), which maintains inventories of observations from all weatherships.

Because of the variety of computers and data formats in use

throughout the world, much effort goes into reading and reformatting exchange data sets.

DATA ARCHIVING

Data are received from the following instruments: Sippican Mk 2a XBT recorders, Sippican Mk8 XBT recorders, Sippican Mk9 XBT recorders, Bathy-Systems XBT recorders, mechanical bathythermographs; serial observations, Nansen water bottles, Neil Brown CTD probes, Bisset-Burman STD/STDSV probes, Navitronics S/V probes and Inter-Ocean CTD probes. The media involved include: hard-copy traces (various scales), cassettes, mini-cartridges, cartridges, 8 in. floppy disc, 5¼ in. floppy disc, seven-track magnetic tape and portable Winchester discs. A processing system has to be developed for each of the instruments, and often for different instruments from one manufacturer, e.g. an internally recording Neil Brown CTD produces different output from an 'up the wire' Neil Brown CTD with an on-board recorder. Each time an instrument or recording medium is changed, a new processing system has to be developed.

Validation

Whatever instrument has been used to make an observation, it is very necessary to validate the results before entering them into the data bank. Great problems arise, especially at analysis stages, if, in striving for a large data set, one incorporates everything, including spurious data. Also XBT observations need particular attention because they are often made by ships on an opportunity basis, by many different kinds of vessel, many of which are not taking the observations for their own use. They are seldom validated and processed by the observer.

On receipt in the data centre, XBT observations are scrutinized for completeness of the records; the ship's position is checked against sea bed depth; time and distance checks are made, and suspected errors are cleared up with the ship if possible.

Evaluation

The traces are then examined to identify any apparent faults in the operation of the probe, launcher or recorder. Last year 33% of observations had identifiable faults and 20% had to be totally rejected. Report-back forms are sent to the vessels which collected the data, detailing the faults identified during evaluation. Vessels are then able to remedy any faults and so improve future observations.

A further validation process is carried out on the 'header' data, using a computer program which checks that the values in each field fall within prescribed limits, e.g. the month must fall between 1 and 12. The program also checks that the positions do not fall on land.

Validation of serial data

The problems with ocean station data are not normally so great, as these observations are made by dedicated ships with scientific personnel on board who will monitor the use of the instrument and carry out basic quality control. However, errors can creep in, and a quality control program is run on all serial data before they are included in the data bank.

Hard copy processing

Acceptable traces are digitized using a line follower on a digitizing table. This produces a stream of XY co-ordinates which is reduced to inflection points by a computer program.

This program works on a fairly simple principle. It tests the first three points on a profile to see if the centre point is more than a set distance from the straight line joining the other two. If it is not it extends the field to include the next point and repeats the test until it finds a point which is more than 0.4 mm (the equivalent of 0.1°C) from the straight line joining the first and last points. It then selects the last but one point as an inflection point. This ensures that all points selected are within the pre-set tolerance. See Figure 2.3.

The same program converts the selected X–Y co-ordinates into depths and temperatures, using the algorithms appropriate to the probe type, and merges the data with the header to produce an XBT record.

Processing digital records

If the observation is recorded on one of the digital media already mentioned, the digital record cannot be accepted without editing. There is extraneous information on the record, e.g. errors in surface values due to thermistor time lag or data recorded after the probe has hit the bottom or the wire has broken.

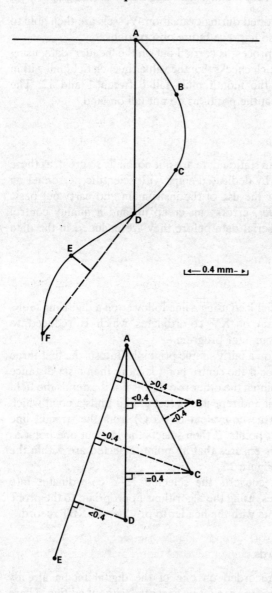

← 0.4 mm →

Figure 2.3 Data compression method
Source: Hydrographic Office

Serial data processing

Probe data are received on magnetic tape and are run through a suite of programs which convert the raw signal into values of depth, temperature and salinity; display the data in a form suitable for manual editing; remove spiking caused by interference, etc.; and combine the data with the header information. As with XBTs, the data can then be reduced to inflection points, though for serial data this is more complex, as inflection points must be identified in both the temperature and the salinity profiles which must then be cross-interpolated to produce temperature and salinity at common depth levels.

Manual digitization is used for observations of a non-standard nature, or, for example, when probe data tapes are damaged and the real-time plot must be used.

Validation of exchange data

The exchange of data on magnetic tape presupposes that the data centre supplying them has applied similar quality control to our own. However, experience has shown that some further computer validation may need to be done. On receipt all observations are run through a program to check their distribution, and the validation program, which is run on in-house processed data, is also run on exchange data to check that all fields fall within prescribed limits and to determine whether the observation falls on land.

DATA USE

Data banking is the easy part! The objectives are clear and programs can be written to massage data into required formats. But there is no point in building up a data bank which is not used. With analysis tasks there is no uniformity; the answers are as many and varied as the questions asked. Requests received in Taunton vary from simple questions that can be answered from work already done or simple enquiries for small print-outs of data to far-reaching questions that require a major analysis program. To meet these demands, which come from both MOD and civilian sources, we have a range of computer programs, first to retrieve the required data sets and then to analyse and display them.

SERIAL DATA (OBSERVATIONS)   HYDROGRAPHIC OFFICE TAUNTON

COUNTRY   74   YEAR 1960   INSTR CODE   SEA BED DEPTH   2000   MARSDEN SQ   217
INSTITUTE   MONTH 06   DATA TYPE   ONE DEG SQ   07
INSTITUTE CD   DAY 02   DATA MODE   0   MAX OBS DEPTH   14
SHIP NO   GMT 1299   DATA METHOD   4   MIN OBS DEPTH   0   LAT 60 00   N
SHIP CODE   0   LONG 07 00   W
STATION   000095   ORIG UNITS 0   ORIG CRUISE 00001960   QUAD (ICES)   1
                                                                                               DETERMINATION 0
WIND DIR 16   DRY BULB   WAVE PER   9   SEA STATE   BAR   ACCURACY   0
WIND SPD 13   WET BULB   WAVE HT   CLD AMOUNT 9   WEATHER 9   RESERVED   0
WATER COLOUR   WATER TRANSPARENCY   SALINITY SCALE

             CORRECTIONS THAT HAVE BEEN APPLIED TO THE ORIGINAL DATA
             DEPTH=   M; TEMP=   C; SAL=   PPT; SV=   M/SEC

COMMENTS ON DATA

                               VALUES FROM OBSERVATIONS

                                               SV   SV   SV   SV   SV
                                             OBSERVED COMPUTED   COMPUTED   UNSPEC'D   GRADIENT
DEPTH QUAL TEMP QUAL SAL QUAL SIGMA-T   (WILSON)   (CHEN &   (X 100)
                                                                  MILLERO)
 0   5   14.00 4   35.412 4 26.523   1504.7   -30.0
 1   5   13.90 4   35.413 4 26.545   1504.4   -30.0
 2   5   13.80 4   35.414 4 26.567   1504.1   -30.0
 3   5   13.70 4   35.415 4 26.588   1503.8   -40.0
 4   5   13.60 4   35.416 4 26.610   1503.4   -30.0
 5   5   13.50 4   35.417 4 26.632   1503.1   -60.0
 6   5   13.00 4   35.418 4 26.735   1501.5   70.0
 7   5   13.50 4   35.419 4 26.633   1503.2   -30.0
 8   5   13.40 4   35.420 4 26.655   1502.9   -40.0
 9   5   13.30 4   35.421 4 26.676   1502.5   -30.0
10   5   13.20 4   35.422 4 26.697   1502.2   -30.0
11   5   13.10 4   35.423 4 26.718   1501.9   -30.0
12   5   13.00 4   35.424 4 26.739   1501.6   80.0
13   5   13.00 4   38.412 4 29.054   1505.4   0.0
14   5   13.00 4   38.422 4 29.061   1505.4

Data retrieval

Software for this must not only be capable of simple tasks as finding all data for a given geographical area or period, but must be capable of more complex finds, e.g. to find all observations for a specific vessel in a defined area during a limited time period.

Data products

The most basic product is a listing of data in its archive format. This is difficult to read, and formatted listings are more usually supplied. These have the added bonus that parameters such as density and sound velocity can be calculated and displayed by the listing program. See Figure 2.4.

Plots of data sets can be produced to show the spatial distribution of observations or instrument type. These are usually produced as overlays to fit Admiralty charts, but line-printer versions are also possible. Plots can also be produced to show the profiles of observations. Observations can be plotted in either half-monthly, monthly, seasonal or annual periods. Plots of depth against temperature, salinity, sound velocity or sigma-*t* or of temperature against salinity can be produced. See Figure 2.5.

The next level of product includes programs which carry out specific tasks on a given data set. For example, we have programs to calculate mixed layer depth and temperature gradients within a data set or perhaps to produce an analysis of maximum and minimum temperatures at standard depths, together with an average temperature profile with standard deviations.

Other programs calculate and plot profiles as a section, with distance, time or an arbitrary stepping point on the *x* axis (Figure 2.6). These 'washing lines' may be based on temperature, salinity, sound velocity or sigma-*t*. The program may also be used to plot partially contoured data values which can subsequently be completed manually to provide a section through the water column.

Representative profiles

A file of representative data is held. This file has been built up by oceanographers at Taunton from analysis of source data for discrete ocean areas.

Profiles for each half-month period in each area analysed so far are stored. However, the task began about five years ago and there are many more areas awaiting analysis. Temperature, salinity and sound velocity profiles are on file, the last having been calculated from temperature

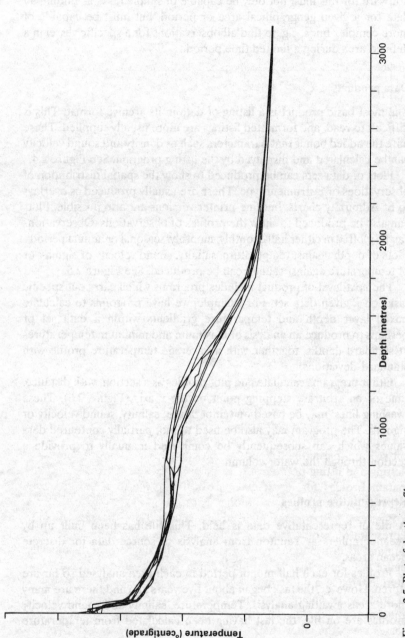

Figure 2.5 Plot of depth profiles
Source: Hydrographic Office

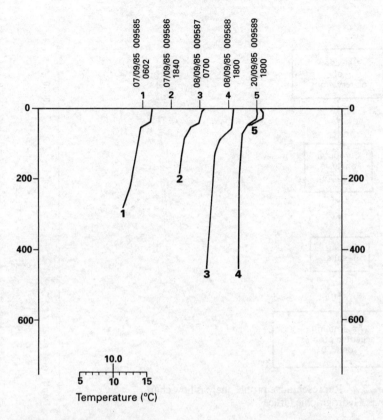

Figure 2.6 'Washing line' plot
Source: Hydrographic Office

and salinity. The file employs the same recording format as that of its source data, permitting the use of the same application programs, e.g. plotting and listing programs. It is planned that this file will eventually contain profiles for about 1000 areas in the North Atlantic, South Atlantic, Mediterranean and Indian Ocean.

Figure 2.7 shows the deceptively simple flow diagram for the representative profile system. The first problem faced is the quantity of data and its distribution in both the spatial and the vertical dimensions. Figure 2.8 shows that as profiles terminate at various depths we are often looking at a different data set for each level. Sparse data sets for an area can be enlarged by increasing the size of the area or going to adjacent time periods. The former method is dangerous where water masses meet; the latter can result in data sets which look good but are

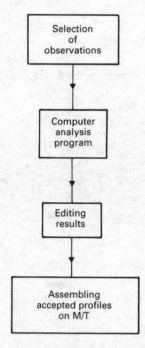

Figure 2.7   Representative profile analysis flow chart
Source: Hydrographic Office

in fact based on only a few observations. The problem does not end there. Data sets are not random. This is especially true of serial data. Oceanographic surveys are mounted to investigate the unusual; there is little interest in pure data gathering, with the result that data come in intensive bursts in particular areas, followed by little or no data.

Once the data sets have been selected they are given to the computer analysis programs which produce representative profiles of temperature and salinity. These profiles are only as good as the data set. The computer is not an oceanographer and often produces profiles which could not exist in reality, e.g. with unstable densities. The results must therefore be edited to produce profiles which are valid not only individually but in relation to other profiles for the same area over the year and in relation to neighbouring areas.

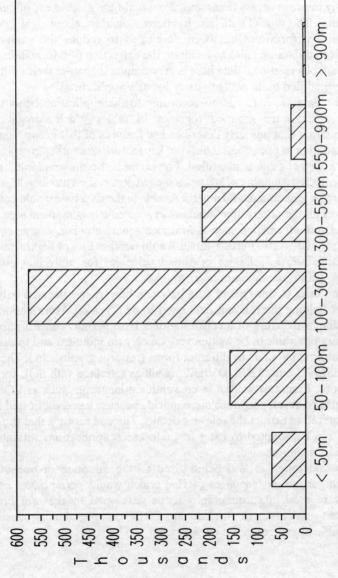

Figure 2.8 Histogram of numbers of observations against depth reached
Source: Hydrographic Office

CURRENT DEVELOPMENTS

Software

Plans are in hand to convert the two main existing data banks to relational data bases. The DBMS planned is the Data General Corporation's implementation of SQL. It is considered that this will greatly improve access times and allow real-time assessment of current holdings. This type of data base has been around for about eight years in its present form and has been developed to reduce the amount of duplicated data and also to facilitate the extraction of information. The concept of a relational data base is very simple. It enables their structure to be modified quite easily to cater for new applications.

The data are not ordered according to the application being used; instead data are grouped together in tables which contain only information that uniquely describes key features of that group. If a key or data item in one group is also the key feature of another group a link with a further table is identified. For example, oceanographic data are referenced by a number of key fields such as observation number, Marsden square, degree square and month. A table in the data base would contain a list of all the observation numbers in a specific geographical area with associated information such as Marsden square number, degree square number, dates, etc. Further tables would contain lists of depths related to temperatures, salinities or sound velocities for individual observations, with observation number as a key.

Enquiries to this type of data base can be very fast because only the tables holding the relevant data need to be accessed. More important, the simple structure of a relational data base permits general-purpose enquiry programs to be written very quickly (in minutes) and to display the information in many different forms (reports, graphs, etc.). The DG 'PRESENT' and 'TRENDVIEW' facilities interface with SQL for this purpose. Programs written in conventional languages such as COBOL can also be developed much more quickly because less complicated logic is required to extract the relevant details. The end result is that the data base can be adapted to cater for unforeseen applications without the need for major change.

Consideration is also being given to the adoption of one of the fourth-generation languages (4GLs) which would permit some of the software used in oceanography to be developed by the user branch without recourse to traditional COBOL programming facilities available elsewhere in the department.

Hardware

To permit improvements in the data storage and handling procedures further enhancements of the MV 8000 computer have been identified: 4 Mbyte additional memory, two 592 Mbyte fixed drives to allow 1 Gbyte for each main data file, a 592 Mbyte fixed drive to provide additional work space, additional disc controllers, four VDTs, 3 interactive graphics work-stations. The interactive graphics work-stations should greatly improve the techniques used for the analysis of oceanographic data, in particular in the determination of representative profiles. They will also play an important part in data archiving, since readers for the various magnetic media outlined earlier will be interfaced to these work-stations, where data profiles can be displayed on the graphic screens for evaluation and editing.

New data

Oceanographic instruments continue to proliferate, and data from Batfish and Thermistor chains as well as new methods of recording data in the Royal Navy are going to influence our activities considerably and demand new solutions. It is often difficult for those whose task it is to acquire and data-bank observations to convince the collectors that they should make their observations available, and design their systems to facilitate data compatibility and exchange. The problem becomes more acute as technology advances. It is now possible to collect huge quantities of data during an oceanographic cruise. Processing systems must be designed and tested before the new instrumentation is deployed or it is likely that the data will never reach a data bank. All organizations involved in the collection of oceanographic data need to be aware of the various uses to which the data may be put; ship time is expensive in comparison with the provision of processing facilities for the data being collected.

3 Oceanographic modelling

John M. Huthnance

This chapter reviews the purposes, scope and present state of oceanographic modelling, with reference to existing and potential management
applications. Particular attention is paid to predictive numerical models.

Forecast models of tides, meteorologically driven surges and waves in
shelf seas are perhaps furthest developed, and are particularly relevant
to coastal protection, coastal and offshore engineering and navigation.
Such models can also be used to improve statistical estimates of extreme
and long-term conditions for design purposes.

Circulation, dispersion and transport modelling is at an earlier stage
of development, and validation data are generally harder to obtain.
Nevertheless, such modelling has important applications to (e.g.)
fisheries, waste disposal, accidental spillage and dredging. These applications also require the integration of biological, chemical and sediment
interactions and transports.

INTRODUCTION

Assumptions about the sea are implicit in every decision regarding its
use. At the most naive level, going for a swim involves expectations: about
sea temperature in relation to the season and location; about sea state
in relation to the weather; increasingly about pollution in relation to
possible sources. Even this example clearly shows the use of conceptual
models, albeit poorly formulated and perhaps only subconscious. It is to
be hoped that decisions involving millions of pounds or national
well-being are based on more conscious judgement using better-
formulated assumptions and models, e.g. as for the Thames Barrier. The
aim here is to encourage thought about models at various levels,
through a review of their scope and limitations, so as to encourage their
deliberate and more effective use.

The essence of any model is that it should represent the real world. For the seas, this should be in a reduced and probably quite different form, which nevertheless accurately epitomizes the question in hand. There is perhaps a growing association of 'model' with 'computer calculation' but it is useful to take a wider view, to lead on from the initial conceptual model and to embrace other forms more appropriate when there are data but there is little theory, for example.

After surveying the range of user interests and model subjects, purposes and types, particular attention is paid to the state of models of sea levels and currents with which the author is most involved. Questions of model quality, validation and transfer to wider use are discussed.

USER INTERESTS AND SUBJECTS

Table 3.1 lists (left) a range of interests in the sea where modelling may be expected to have a role. Typically, this is because information is important to obtain (for economic or health and safety reasons, say) but is not directly available (because it pertains to the future or because measurements are impractical, more expensive or less informative than modelling, say). On the right in Table 3.1 with an attempt at rough correspondence or relevance to the user interests, are subjects of potential or present oceanographic modelling.

Many of the interests in Table 3.1 may well be reinforced by legislation. On the other hand, oceanography is a developing science rather than an accomplished art, and the greatest use of models is probably by scientists themselves, with interests among the subjects on the right of Table 3.1. In similar vein, recent developments in measurement techniques (especially by remote sensing) involve models, notably of radio/radar–sea interactions. For any one application, the reduction involved in a model is likely to imply specialization to just one or a few of the subjects in Table 3.1, and according to context: ocean basin, continental slope, shelf, estuary, river, intertidal zone, lagoon, marsh or beach.

MODEL PURPOSE AND TYPES

Other aspects of models are their purpose (in functional rather than topical terms; Table 3.2, left) and their intrinsic type (Table 3.2, right).

The correspondence between purpose and type in Table 3.2 is not clear-cut, and any particular model may qualify under several types and fulfil several purposes. A fuller description of each type follows,

Table 3.1 Maritime interests in which modelling has a role

User interests	Oceanographic modelling subjects		
Navigation, sailing, traffic routeing	Wave height, icebergs		
Operations, construction, surveying diving, safety	Wind forcing, currents		
Search and rescue			
Coastal defence/flooding	Tsunami		
Shoreline definition	Sea levels trends, geodesy	Tides	Surges
Coastal morphology	Wave propagation, inshore dissipation, breaking		
Beach nourishment	Sediment; suspended and bedload		
Dredging	Erosion, deposition, sorting		
Engineering design: bed stability (cable, pipe, pile, platform, ship mooring, riser, barrier, harbour tidal/wave power)	Bed structure, strength		
	Wave spectra		
	Current profile with waves		
	Wave diffraction		
OTEC	Water masses, convection		
Signalling – submarine detection	Temperature, upwelling, internal/inertial waves and tides		
Exploration – seismics	Acoustics		
Extraction – gas/oil/gravel/mining			
Spoil, dumping, spills	Transport and dispersion: circulation, diffusion, distributions in 3D and time eddies/fronts/plumes/mixing		
Amenity/pollution			
Radioactive waste	Benthic exchange, pore water		
	Sediments: absorption, bioturbation, flocculation erosion, suspension		
Sewage sludge			
Water quality and oxygen			
Metals			
Toxins	Air/sea exchange: surface film, bubbles		Chemistry
Heat	Temperature		
Disease, stress	Salinity, blooms, eutrophy		
Fishery	Nutrients, upwelling		
Aquaculture	Light, turbidity, turbulence	Biology	
Conservation	Production, food webs, stocks		
Ecological/environmental assessment	Biogeochemical cycles, budgets		
Acid rain	Biogenic (DMS)		
Weather and climate	ENSO, palaeoceans, ice		
Monsoons, hurricanes	Heat transport		

Table 3.2 Models: purpose and type

Purpose	Type
Organize/document knowledge	Organizational
	Empirical, spectrum
Communication	Conceptual
Reveal issues (for research, observation)	
Link cause and effect	Physical
Experiment (test/improve understanding)	Analytic
	Semi-analytic
Address questions, 'What if ...?' (scenarios, results of changed forcing/context/ parameters)	Numerical
Prediction evolution	Prognostic
Forecasts (perhaps routine)	Operational
Test models and strategy	Hindcast
Experiment design	
Survey/monitor strategy	
Improve statistics	
Extremes, distributions	Statistical

including illustrations from the well developed field of storm surges and circumstances, where it may be most useful.

An *organizational* model may often be a block diagram outlining a modelling or experimental strategy, e.g. Figure 3.1. An updated diagram for the scheme run operationally at the UK Meteorological Office is given in Heaps (1983).

An *empirical* model might be regarded as one form of organizational model in that observations are arranged by variable type. Variables regarded as dependent are related to those regarded as independent, e.g. (Townsend, 1981):

Tyne surge $\approx a$ (Wick surge) at time $t-TW + b$ (320° wind at Fair Isle) at time $t-TF + ...$

where the coefficients $a, b ...$ and time delays $TW, TF ...$ are chosen to minimize the errors over the available data. Such a model is useful for particular forecasts where good data are available, but theory is less well developed or does not furnish a means of practical or sufficiently accurate prediction. A disadvantage may be that coefficients derived from many typical occurrences do not apply well to the few extreme events which are of most concern. Tidal predictions are the best developed empirical models by far, illustrating the benefits to be

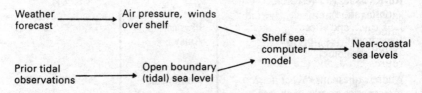

Figure 3.1   A block diagram as organizational model

obtained when theoretical knowledge allows a judicious choice of independent variables (*viz.* constituents of the tide-generating potential). Principal component analysis derives relationships between variables not distinguished as 'dependent' or 'independent' (Kundu *et al.* 1975).

The Garrett–Munk (1979) spectrum for oceanic internal waves and the Pierson–Moskowitz (1964) or JONSWAP (Hasselmann *et al.* 1973) spectra for surface waves are also empirical models where spectral forms have been fitted to observations by adjusting parameters.

A *conceptual* model expresses, usually in words, our understanding of the sea's behaviour, e.g. 'A northerly wind drives a southward flow (trapped against the east coast of Great Britain by the Coriolis force) so that levels are raised by water arriving in the southern bight of the North Sea.' Such a model typically states which factors are thought to be involved, and may be represented by a block diagram (not necessarily to any great benefit). It is rarely quantitative, but is often the most effective means of communication, for the scientist and 'lay' person alike.

Physical, analytic and *numerical* models represent the real world by an idealized scale model, by mathematical equations and by (usually approximate) numerical realizations of such equations. All imply a prior conceptual model and are thereby limited in scope or faithfulness to reality. (For a physical model this may not be obvious, but any lack of a conceptual model implies a corresponding lack of confidence in the idealized, scaled results. An estuary model, for example, may be correctly scaled according to hydraulic concepts but then have the wrong friction and mixing for determining the salinity distribution.) They have the advantages of quantitative results and ease of experimentation with altered conditions, and reduce dependence on observational data.

The terms 'analytical' and 'numerical' implicitly suppose that a solution to the equations is available. For an analytical model, this represents a restriction on the complexity of the equations and especially on the domain where a solution is sought. Usually only unrealistically idealized domains are tractable. Numerical models typically divide a complex domain approximately into small discrete portions of simple form (e.g. cuboids in three-dimensional space); each portion as a whole satisfies the (numerically realized) equations, usually involving adjacent values so that the portions evolve jointly. Conscious decisions are required regarding the number of spatial dimensions (and time?) to be represented, and the use of appropriate data or conditions at boundaries where adjacent values are not inherent in the model. Analytical and numerical models may use reduced equations (e.g. kinematic rather than dynamic, hydrostatic or geostrophic, omitting relative acceleration) for simplicity or to assess the importance of omitted effects. For example:

$$\delta\zeta/\delta x, \; \delta\zeta/\delta y = \underset{\sim}{\tau}/\rho g h$$

determines sea level slopes directly from gravitational acceleration, g, sea depth, h, and density, ρ, and the wind stress, $\underset{\sim}{\tau}$. Explicit solution is easy for a wide range of algebraic forms of $\underset{\sim}{\tau}/\rho g h$, provided that its integral around any closed curve is zero (one condition for validity of this reduced equation, which omits all inertia and friction). Under the same condition, numerical solution can relate all values ζ on an x–y grid back along a combination of x–steps, Δx and y–steps, Δy, to a given value at one location, by representing:

$$\delta\zeta/\delta x = (\Delta x)^{-1} \left(\zeta(x + \Delta x, y) - \zeta(x, y) \right)$$

(and similarly for y–steps) and using an average value for $\underset{\sim}{\tau}/\rho g h$ along each step.

A *semi–analytical* model may be analytical in spirit but require numerical evaluation of a complex but otherwise explicit solution. Alternatively, analysis may be used to reduce the dimensions treated numerically, e.g. by supposing sinusoidal dependence (or uniformity) in time or along a continental shelf of uniform cross-section. Wave calculations by ray tracing use a simplifying assumption:

wavelength < < topographic length scale

to reduce a fully two-dimensional problem to many separate one-dimensional integrations for ray paths and wave amplitude evolution along them (Berkhoff *et al.* 1982). Analytical and semi-analytical models are valuable for providing insight and independent solutions with which to interpret and test the numerical models required for complex realistic calculations.

A *prognostic* model (or equation for one component thereof) relates time rates of change (for that component) to contemporary values and thereby predicts to future time. This property is evidently required for forecasting and predicting evolution, and (if present) will appear explicitly in the formulation of all model types above, except that it may be intrinsic to a physical model. Thus:

$$\delta\zeta/\delta t = -\delta(uh)/\delta x - \delta(vh)/\delta y$$

predicts that sea level will be higher at a later time t if the transport h x velocity (u, v) is at present converging.

An *operational* model is well proven and gives results, to those who require them, sufficiently in advance of the time to which they pertain. The distinction is made because most oceanographic models are not operational, nor need they be. This property is evidently required for forecasting, and entails a prognostic model. However, the main problems are likely to be the organization of data input to the model and the rapid dissemination of results. Weather forecasts are a well known example, but wave and surge forecasts are also carried out by operational models for UK waters and beyond at the Meteorological Office (Golding 1985; Heaps 1983). Difficulties of model operation *per se* may be circumvented by holding the results from prior runs of a sufficient range of scenarios, an approach commonly adopted for wave predictions near shore (BSI 1984).

A *hindcast* model is run as though carrying out a forecast but after the event, and perhaps with better input data than would have been available for a real forecast. Such calculations are common to test models and critical inputs to them, and hence for designing experiments and surveys or monitoring schemes. If a (deterministic) model is well established (for wave predictions, say, on the basis of meteorological data), then it may be used to improve the statistics of a sparsely observed output variable (waves) on the basis of better known statistics for the input variables (winds). Such is the aim of the NESS project (North European Storm Study, Francis 1987). Evidently this approach presumes calibration and validity of the model for the possibly extreme events of most concern.

A *statistical* model describes the probability distribution of values of (say) sea level at one port, on the basis of previous observations but possibly without much other understanding, knowledge or data. This is a distinction from empirical models, although hybrids can be conceived and (e.g.) knowledge of the tidal contribution to total sea level enables better use of limited observations in determining the probability distribution (Pugh and Vassie 1980). Statistical models are important for estimating extreme values and variances direct from observed data. A special form of statistical numerical model simulates dispersion in a given velocity field by 'tracking' particles; diffusion is modelled by additional random particle movements. Constituent distributions are obtained (most effectively near a source) by aggregating results for perhaps many thousands of particles.

No one 'best' type is likely. Storm surge forecasts are carried out routinely at the UK Meteorological Office by an operational, prognostic numerical model. This incorporates the north-west European shelf-sea geometry and a full range of our concepts for long waves: two (horizontal) dimensions, non-linearity with tide–surge interaction, especially in the bottom drag on the flow, and direct input of computer-forecast meteorological data (Heaps 1983). Nevertheless, organizational models play a useful role in its description and possible further development (to incorporate observed surge data or a rearranged sequence of analysed and forecast meteorological data, for example). Empirical models for individual locations, once the principal means of forecasts, are still run alongside the numerical model and may provide a partial but useful check (against corruption of computer files, for example). Statistical models are still used to provide estimates of extremes where data are available at individual ports, more directly and conveniently than by model simulation of many storms, say.

QUALITY

A numerical model is usually only an approximate realization of the governing equations. However, a choice of approximations is usually available, and emphasis may be placed on faithfulness in various aspects. Thus numerical schemes may be designed to conserve mean-square vorticity (in ocean models, inhibiting grid-scale 'noise') energy (bounding large-scale motion) or a passive constituent (say) but possibly at the expense of unrealistic diffusion. This is the subject of many numerical fluid dynamics texts, and details are beyond our present scope. However, the choices imply a need for priorities in which aspects of the model calculation should be most accurate. Satisfaction of conservation properties is desirable in general terms but has less relevance in shallow shelf seas dominated by friction and mixing processes.

Friction and mixing are primarily the result of turbulence and eddies, and are inherently stochastic. In the deep ocean the spatial scale of such stochastic behaviour is greater, including eddies of (internal Rossby) radius 50 km generated by instabilities in the Gulf Stream, for example (Robinson 1987). Predictability is then limited to the growth time of such features from first observational detectability to significance (~ 10–100 days) and emphasis is placed on the ability to assimilate constraining up-to-date observations into a prediction model (Robinson 1987). There is a close parallel with weather forecasts, in which most cyclones (depressions, scale 1000 km) are stochastic (growth time ~ 10 days) and are predicted individually only for evolution after being observed.

On larger spatial scales – perhaps 20 km or more in shelf seas and 100 km or more in deep oceans – only statistical aggregates of the stochastic features may be significant, e.g. turbulent friction on the 'mean' flow, which is usually regarded as deterministic. Numerical models then require a choice between (1) a coarse resolution unable to represent the stochastic features and (2) their proper representation through fine resolution in 'eddy-resolving' models. Such resolution (2) involves numerical computation of a scale which is now becoming feasible on the fastest computers, for ocean basins and for ~ 5 km radius eddies in shelf seas. Confidence is required that the model physics correctly aggregates the stochastic features despite a large measure of random cancellation. Similarly, coarse resolution (1) requires a 'parameterization' entailing some independent information or model to represent aggregate effects of the stochastic features. Typically, such parameterization (bottom drag coefficient in a surge model, say) are an oversimplification. Values satisfactory for one context may in effect represent a 'tuning' of the model there, and it may fail if used elsewhere. Tests for (lack of) sensitivity to uncertain parameterizations or values are important for confidence in approach (1).

Compromises, as in (1) versus (2) above between simplicity and accuracy, or extent and resolution, are inherent in the reduction which characterizes any model in comparison with the prototype. Grid resolution in a numerical model is always a consideration, computing capacity being finite (a factor of two in horizontal resolution becomes a factor of eight in computation, for explicit time integration of the shallow-water equations). Sub-grid processes must be parameterized whether or not they are stochastic. The analogue of many wave models is neglect of diffractive variations of amplitude in one or both horizontal dimensions (giving parabolic and ray-tracing models respectively: Berkhoff *et al.* 1982) so that the wavelength need not be resolved. Omission of processes also reduces a model and the associated computation, but is usually motivated by the desire for simplicity, understanding and confidence in model operation (rather than results). Sophistication may eventually benefit simulations, but only in so far as all aspects are tested and validated.

VALIDATION

The progress to well developed surge models (Heaps 1983) is instructive. The topic has attracted for decades a large number of analytical semi-analytical and laboratory studies for various idealized contexts. Amongst these is a 'North Sea rectangle' of fixed grid and dimensions, and known wind-forced solution, on which any proposed (e.g. three-dimensional) numerical formulation can be tested. Thus a large body of well verified comparison data is available and has been used for surge model validation. In a realistic shelf-sea context, tides are usually important with similar dynamics, and extensive accurate data are readily available to test any reasonably large model (say, quarter-wavelength or more, e.g. Lynch and Gray 1987 *et seq.*). If friction is important, then the surge model should simulate tides anyway, to generate the correct level of non-linear friction. Hence model friction is also tested by tidal decay and phase progression. Detailed comparisons with levels at many UK coastal tide gauges for many notable surges have served to verify the operational model.

Such thorough validation has been possible because coastal sea levels resulting from long waves are relatively easy to record and model. Detailed comparisons with tidal currents of a model representing their vertical structure have recently been published (Davies 1987) for homogenous UK shelf seas. However, for the vertical structure of wind-driven currents in various contexts (depth, tidal current strength, stratification, waves) thorough model validation awaits measurements and analysis. Other present lines of development, to combine wave currents and wave spectrum forecasting interactively with tide–surge models, may benefit from considerable work with physical (flume) models (e.g. references in Christofferson 1982). However, there are few observational data to validate the model formulation, although wave–tide/surge interaction is believed to be a significant factor in coastal flooding risk (Alcock 1984). Thus validation becomes increasingly difficult or tenuous for sophisticated models requiring more data and particular combinations thereof.

Development of a water quality model, as in the North Sea Project (NERC 1987), is a case in point. Existing knowledge of the North Sea enables us to identify some important factors *a priori:* a strong but variable seasonal cycle; affinities to abundant fine cohesive sediments; distinctive mixing at frontal boundaries to coastal or seasonally stratified waters; varied sources of metals and other constituents. Thus a need for more concerted data is apparent: to define one complete seasonal cycle for a range of interacting physical, chemical, biological and sedimentological variables; process studies within this context for data on three-dimensional advective currents, diffusion parameters in several distinctive areas, stresses suspending sediments, particulate–water– sediment– and air–sea fluxes, and a range of non-conservative chemical and biological interactions.

DEVELOPMENT AND TRANSFER

Experience with surge models (Heaps 1983) has shown that progress from model conception to routine use may take decades. A long period is likely if data are required to assist model formulation or validation, if numerical modelling techniques have to be acquired, if the model has to be made operational (by organizing rapid and routine inputs and the dissemination of results) or if it has to win wide acceptance, perhaps through extended trials.

A modular approach to numerical model development relates well to computer program structure and testing, and to any organizational model. It may also be natural to the topic, as in the North Sea Project. Any constituent contributing to water quality changes according to: sources or sinks; net convergence or divergence of transports. Transports in turn comprise (1) any (downward) settling velocity for suspended particulates and sediments (2) 'bodily' advection by currents or bed-load sediment transport (3) net diffusion by a molecular or other mixing process, e.g. turbulence, bioturbation. Because (2) and (3) are common to all constituents which do not themselves affect density and hence (2) and (3), there is a natural modelling sequence: domain, forcing, currents and stresses, sediment movement, dispersion, and finally constituent evolution: only at this stage are constituents distinguished by (1) and sources or sinks. The stages are naturally separated as modules (ending with one for each constituent) within an overall framework. Then upgraded modules may be substituted for previously proved or trivial forms as development proceeds.

The framework is in essence an organizational model. Such an approach facilitates: robustness in the model – individual modules and successive combinations can be tested; documentation; flexibility – through a structure for substitutions or extensions. These are all aspects of 'user-friendliness' important to wider model use. They require particular consideration in view of models' usual origins in scientific research, where wider use is not an initial concern.

Successive degrees of transfer for wider use involve model results (perhaps for a range of scenarios), some description of functionality (e.g. an operational or conceptual model, the usual level in publications) or the complete computer code for a numerical model. Complete or detailed transfer requires considerable understanding by the recipient of the model's operation and limitations, and the availability of appropriate input data and computer facilities. (The latter are a receding constraint for many potential recipients.) For such complete transfers (relatively rare hitherto) the practice at the Proudman Oceanographic Laboratory has been to encourage a visit by the user (for some weeks or months according to model complexity) so as to gain expertise in model operation by working alongside the originators.

The UK storm surge model has been transferred to the Storm Tide Warning Service (located at the Meteorological Office) for routine operation. However, it is noteworthy that the Proudman Oceanographic Laboratory, as the originator, has a continuing role in maintaining the model (as the computer or meteorological forecast format is changed, say). Thus transfer has not been complete.

A final consideration in model utilization is any gap between natural model subjects (on the right in Table 3.1) and user interests (Table 3.1, left). Such gaps have to be bridged by consideration of the factors relevant to any interest and a scheme (or socio-economic model?) to assess their impact. Determining the relevant marine factors entails a continuing dialogue between the oceanographer and the user of his model results.

REFERENCES

Alcock, G.A. (1984) *'Parameterising extreme still water levels and waves in design level studies'*, Institute of Oceanographic Sciences report 183, London: IOS.

Berkhoff, J.C.W., Booy, N., and Radder, A.C. (1982) 'Verification of numerical wave propagation models for simple harmonic linear water waves', *Coastal Eng.* 6: 255–79.

BSI (1984) *BS 6349: British Standard Code of Practice for Maritime Structures*, Part 1, *General Criteria* London: British Standards Institution.

Christofferson, J.B. (1982) 'Current depth refraction of dissipative water waves', Tech. Univ. Denmark, Inst. Hydrodyn, and Hydraul. Eng. Series 30, Lyngby: Technical University of Denmark.

Davies, A.M. (1987) *A numerical simulation of tidal currents on the United Kingdom continental shelf*, Department of Energy, OTH 87 271, London: HMSO.

Francis, P.E. (1987) 'The North European Storm Study (NESS)', in Society for Underwater Technology (ed.), *Modelling the offshore environment*, Advances in Underwater Technology, Ocean Science and Offshore Engineering 12, London: Graham & Trotman, pp. 61–70.

Garrett, C.J.R. and Munk, W.H. (1979) 'Internal waves in the ocean', *Ann. Rev. Fluid Mech.* 11: 339–69.

GESAMP (1988) *Review of shelf sea water quality modelling*, Report of Group 25, in preparation.

Golding, B. (1985) 'The UK Meteorological Office operational wave model (BMO)', in SWAMP Group (ed.) *Ocean wave modelling*, New York: Plenum Press, pp. 215–19.

Hasselmann, K., *et al.* (1973) 'Measurements of wind–wave growth and swell decay during the Joint North Sea Wave Project (JONSWAP)', *Deut. Hydrogr. Zeit.*, Supplement A12.

Heaps, N.S. (1983) 'Storm surges, 1967–1982', *Geophys. J. Roy. Astr. Soc.* 74: 331–76.

Kundu, P.K., Allen, J.S., and Smith, R.L. (1975) 'Model decomposition of the velocity field near the Oregon coast', *J. Phys. Oceanogr.* 5: 683–704.

Lynch, D.R., and Gray, W.G. (1987) 'Tidal flow forum – introduction', *Adv. Water Resources* 10: 114.

NERC (1987) 'North Sea Project', *NERC News* 3: 10–12, 4: 10–11.

Pierson, W.J., and Moskowitz, L. (1964) 'A proposed spectral form for fully developed wind seas based on the similarity theory of S.A. Kitaigorodskii', *J. Geophys. Res.* 69: 5181–90.

Pugh, D.T., and Vassie, J.M. (1980) 'Applications of the joint probability method for extreme sea level computations', *Proc. Inst. Civ. Eng.* 69: 959–75.

Robinson, A.R. (1987) 'Predicting open ocean currents, fronts and eddies', in J.C.J. Nihoul and B.M. Jamart (eds) *Three-dimensional models of marine and estuarine dynamics*, Amsterdam: Elsevier, pp. 89–111.

Society for Underwater Technology (1987) *Modelling the offshore environment*, Advances in Underwater Technology, Ocean Science and Offshore Engineering 12, London: Graham & Trotman.

Townsend, J. (1981) 'Storm surges and their forecasting', in D.H. Peregrine (ed.) *Floods due to high winds and tides*, London: Academic Press.

Part II

The management of individual uses: biological resources and environment

4 Marine fish farming in the United Kingdom
A challenge for the 1990s

A.G. Hopper and M.J.S. Gillespie

INTRODUCTION

The farming of fish and shellfish in UK marine waters is now a substantial and rapidly expanding industry whose commercial origins go back only to the late 1960s, when fish cages first started to appear in sheltered Scottish sea lochs. Salmon form the main output by value and weight. Production in 1987 exceeded 12 000 tonnes, with a first sale value of over £50 million. There are some 190 sites contributing to this output, located mainly on the Scottish west coast and offshore islands such as the Hebrides to the west and Orkneys and Shetlands to the north.

Although various marine species have been considered for cultivation over the same period, only turbot is in commercial production at one location, Hunterston in Ayrshire, where 80 tonnes per annum, valued at £0.45 million, are currently produced. The turbot are grown-on from larvae provided by the Sea Fish Unit at Ardtoe in water heated by the power station coolant to 6°C above ambient. This limited development compared with salmon reflects the greater difficulty of rearing marine species through the early larval stages and also the requirement of higher than ambient UK temperatures for economic growth of many of the higher-value species such as turbot, Dover sole and bass. Hopes of future commercial growing alongside salmon are pinned largely on halibut, which in 1987 was reared artificially for the first time in the UK by the Seafish team. Regrettably a fire at the hatchery destroyed the first juveniles. Halibut are expected to take three years to reach marketable size.

While all types of finfish farming require substantial initial capital for equipment and facilities, shellfish cultivation can be undertaken with lower levels of investment. This has led to a predominance of small shellfish production units encouraged by local funding agencies such as

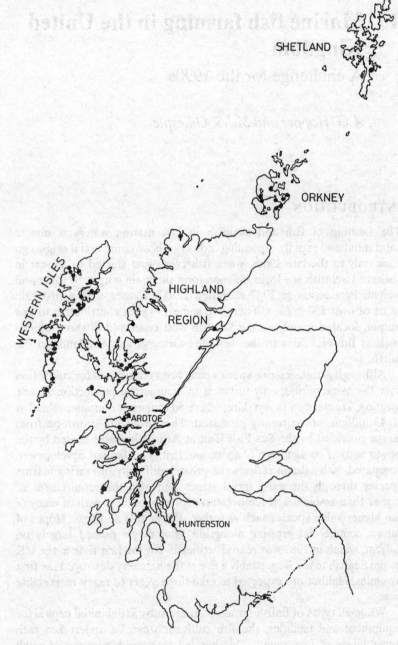

Figure 4.1   Approximate position of growers

the Highlands and Islands Development Board who wish to encourage crofter-type operations by providing supplementary income for the operators. Perhaps as a result of this fragmented approach, development has again been much less dramatic than with salmon. Larger companies have more recently, however, been attracted to take up production of oysters, mussels and scallops. The 1987 Scottish production of farmed shellfish was valued at almost £1 million, and more rapid expansion is now taking place.

The Marine Farming Unit of the Sea Fish Industry Authority (Seafish) is based at Ardtoe in Argyllshire and has been involved in much of the R&D work on new species and methods now being taken up by industry. Although work on salmon is excluded from the authority's brief, the Ardtoe unit is the centre for the development of halibut cultivation in the UK and has been instrumental in introducing scallop cultivation to our shores.

This chapter outlines the current situation in the marine (i.e. non-salmonid) sectors of the industry and sets out some pointers for future development.

SPECIES SELECTION

There are several species of which there is sufficient knowledge to make farming possible. Commercial and practical judgements still have to be made by the investors and their backers, acting on the advice of such experts as the Seafish team.

First, there is little point in farming a species of which there are abundant wild stocks, especially if the availability of those stocks can be expected to continue. Second, farmed fish and shellfish carry certain cost penalties associated with the risk. In the case of finfish the cost of feed and labour amounts to up to 50 per cent of the production costs, and has to be offset by the selling price. Only high-value species are likely to be commercially worth while. Third, there has to be some reasonable expectation of success, both to cultivate a good-quality product and to market it successfully. Small-scale growers can be vulnerable in the market place, especially if they are located well away from the main centres of population.

At present the following are the species which may be considered candidates for marine farming in the UK: oysters, mussels, scallops, queen scallops, manila clams, lobsters, halibut, turbot, Dover sole, cod, bass, eels.

SHELLFISH

Species

The take-up of marine shellfish cultivation is much more advanced than that of marine finfish. The technology is much simpler, the costs are low and the chances of success are higher. Shellfish cultivation is suitable for the small-scale farmer or crofter seeking a supplementary income.

Oyster growing is well established in England and Wales, and production from Scotland and Northern Ireland is increasing. The native oyster, *Ostrea edulis*, is still available from the wild along the south and east coasts of England, although stocks have been decimated by bonamia disease in recent years. It is a slow-growing species (four to five years), which is a disadvantage in the northern part of the British Isles, where growers have a preference for the Pacific oyster, *Crassostrea gigas*, an introduced species which reaches market size from hatchery seed in two to three years. *Gigas* is not susceptible to bonamia but is sensitive to pollution from tributyl tin oxide (TBT), a compound which before being banned in 1987 was the active ingredient in many marine anti-fouling paint formulations.

Mussels are still available in quantity from wild stocks in the Wash, Menai Strait, Morecambe Bay, Dornoch Firth and South Wales, and these areas will supply the bulk of supplies for many years. Semi-cultivation in which natural seed mussels are relaid in deeper water is taking place in the Wash and the Menai Straits. Relaying gives a better growth rate and increases production, since the mussels are less exposed to weather conditions, including extreme low temperatures. Scottish cultivated mussels are generally grown on ropes suspended from surface rafts or floats moored in deep water. This system encourages rapid growth to market size, resulting in high-quality grit-free produce with thin, attractive shells and a value up to five times that of the wild stock. The Seafish team at Ardtoe are currently investigating the best densities to balance growth against costs.

A method already well established in France has been tried with success in South Wales. This is *bouchot* culture, in which seed mussels are placed in netting tubes and wrapped around poles on intertidal beds. Excellent growth and quality mussels are obtained.

Scallops, *Pecten maximus*, are found as wild stock on the English south coast, in the Irish Sea, the Clyde and the Scottish north and west coasts. Opinion is divided as to whether or not the world stocks are in decline. It is clear that overfishing reduces certain beds to a level below which they can no longer be economically exploited and fishing has to

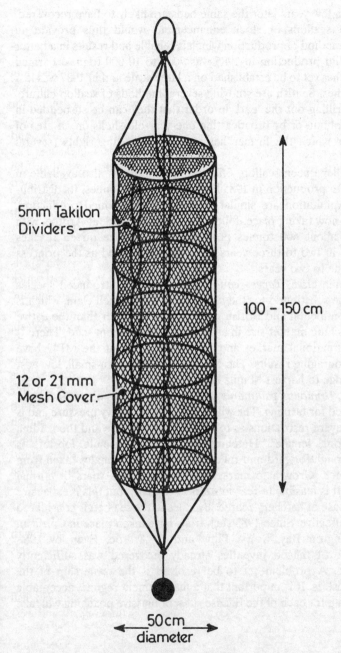

5mm Takilon
Dividers

12 or 21 mm
Mesh Cover.

100 – 150 cm

50 cm
diameter

Figure 4.2 Scallop culture lantern net (seven or ten level)

cease, and a few years later the same beds are likely to have recovered. Farming of scallops or stock enhancement would thus provide an alternative method of production which is reliable and results in a better product. Wild production in 1985 was 7000 to 10 000 tonnes. Farmed production has yet to be established on a large scale and in 1987 was less than 10 tonnes. Seafish are studying several methods of scallop culture, including drilling out the 'earl' in order that they can be suspended in the water column or by broadcasting one-year-old shells on an area of sea bed for which the farmer has secured harvesting rights (several orders[1]).

The smaller queen scallop, *Chlamys opercularis*, is also available in the wild, with production in 1985 from 6000 to 9000 tonnes. Its distribution and exploitation are similar to the scallop. Farmed production, however, is now taking place, following successful work at Ardtoe, with production about 600 tonnes per year. The full-size queen reaches market size in two to three years but can be marketed as the princess scallop in one to two years.

The Manila clam, *Tapes semi-descussata*, is an introduced species which can be substituted for the *palourde* or carpet-shell clam, which is found in Continental and British waters. It grows faster than the native species, reaching market size in around three years from seed. There is a good international market and production trials at the MFU have yielded encouraging results. The MFU has developed a small, low-cost suction dredge to harvest Manila clams.

Lobsters, *Homanus gammarus*, are the only species of crustacean to be considered for farming. The wild stock is under heavy pressure and is subject to severe restrictions as to landing seasons, size and the landing of egg-bearing females. Hatching and rearing juvenile lobsters is relatively straightforward and relies on egg-bearing females taken from the wild stock. Growth to market size is five to six years in normal conditions. It is feasible to rear lobsters in captivity, but this is expensive and the release of hatchery-reared juveniles to the sea bed is considered more cost-effective. Since 1983 such trials have taken place in Cardigan Bay, Bridlington Bay, Scapa Flow and at Ardtoe. Even by 1988 the number of tagged juveniles already recovered was sufficiently encouraging. A problem yet to be resolved is the ownership of the released lobsters. It is important that a management regime, acceptable to all, is set up for each of the release sites, or massive poaching will take place.

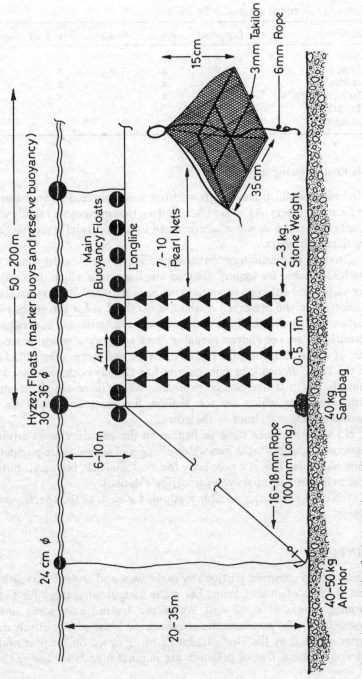

Figure 4.3 Longline for intermediate scallop culture

Table 4.1 Production methods for shellfish

Shellfish	Trays	Longlines	Rafts	Bouchot	Broadcast	Lagoons
Oysters	x	x	x			x
Mussels		x	x	x	x	x
Scallops		x	x		x	
Queen scallops		x	x		x	
Manila slams					x	
Lobsters		x				

Shellfish growing methods

Molluscan shellfish may be grown from seed collected either naturally or from a hatchery. As all are filter feeders, they grow more rapidly when totally immersed in water where there is a good tidal interchange of nutrient-rich water.

Growing in plastic trays either intertidally or in subtidal waters is a method suitable for oysters. Buoyed longlines from which the shellfish are suspended either on ropes or in net bags are suitable for mussels and scallops, and the preferred method in Scotland is for the ropes to be suspended from rafts. The broadcasting of seed on the sea bed, either in an enclosed area or under a predator-proof net, is suitable for all species but, of course, increases the cost of recovery, and there is the added risk to the stock. *Bouchot* or pole culture has already been mentioned for mussels and it is likely to be profitable where there are sheltered or estuarine areas which are too shallow for boats and there are no environmental objections to the poles.

No work has been done in Britain on the construction of artificial lagoons or onshore tanks into which changes of seawater are pumped at each high tide. This is a possibility for most shellfish, but needs further research into cost-effective engineering solutions.

Table 4.1 shows the possible methods for each of the species under review.

FINFISH

The Scottish coastline, particularly in the west and around the northern and western offshore islands, has many natural advantages for finfish farming. Areas of south-west Wales and Ireland have some similar possibilities. The requirements are sheltered areas with relatively deep water warmed by the Gulf Stream and free of ice build-up in winter. Obviously, areas free of pollution are important and, as most of these areas are located in rural parts of the British Isles, they receive either government or EC regional aid, a considerable financial inducement for potential growers.

The basic methodology of hatching, rearing and on-growing turbot (*Scopthalmus maximus*) has been understood since the mid 1970s. Although on-growing is possible in ambient waters, warm water at about 17°C is essential for commercial success. For this reason the only on-growing facility in the UK is at Hunterston power station on the Ayrshire coast. A similar facility is situated at Gravelines in France. In north-west Spain, however, the ambient temperatures are ideal for turbot cultivation, and it is there that major production expansion is likely to take place.

Turbot hatcheries exist at Hunterston and Ardtoe and another commercial hatchery is planned for the Isle of Man. The turbot hatchery system is based on manual stripping of eggs from broodstock which come into spawning condition once every year. The broodstock can be induced to spawn at different times of year by manipulation of artificial daylight to give an all-year-round supply of juveniles.

Stripped eggs are fertilized and incubated for about six days before hatching. The hatched larvae live on their yolk sacs for two days and then require their first added nutrition. This is critical, and the feeds are based on cultured rotifers and larval crustaceans. At forty days the juveniles can be weaned on to a pelletized diet. For on-growing a good-quality industrial fish diet, such as sprats or sand eels, is provided. Growing time to 2 kg in a temperature range of 15–18°C is about two and a half years, which is about half the time in ambient water conditions found in the UK.

Halibut *Hippoglossus hippoglossus*, is still in the early stages of development, but has the advantage over turbot that it can be reared in UK ambient water temperatures. Research into halibut culture is continuing at MFU Ardtoe and in Norway, Denmark, Iceland and the Faroes. The MFU achieved a considerable breakthrough in 1987 with the successful rearing through the metamorphosis stage of six juveniles, although they were destroyed in a fire early in 1988.

The process of stripping, hatching and on-growing is similar to that for turbot, but so far the spawning condition takes place only between January and April. It is expected it will be three to five years before halibut is commercially farmed in Britain.

It has been known for some time that cod, *Gadus morhua*, can be cultivated using similar methods to those for turbot, but without the need for high on-growing temperatures. It is only because this species is severely depleted in the wild that interest now exists in farming. Over

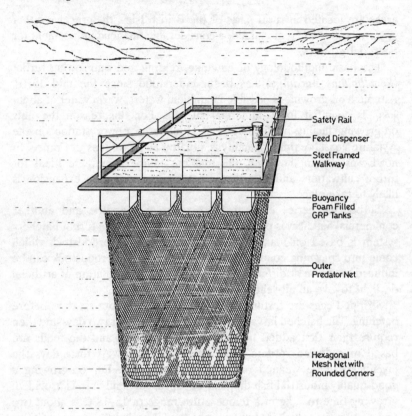

Figure 4.4   Typical sea cage

the last five years Norwegian scientists have developed a large-scale lagoon system for the early stages of rearing cod to a suitable size for further on-growing in sea cages. Such a system could be adapted to Scotland and western Ireland, although it is not cost-effective at present.

Cod can be grown in salmon-type cages. There is a tendency to cannibalism which might be overcome through grading the stock frequently and removing the larger fish.

In the past there have been a number of attempts at developing commercial fish farms for eels, *Anguilla anguilla*, again using waste heat from power stations or other industrial processes. The only current farm is at Hinkley Point power station on the Severn. A difficulty with eels is that the domestic demand is small and the main overseas opportunity is in Holland and Germany. Market links would have to be established to ensure a viable operation.

THE NEXT STAGES IN FISH FARM DEVELOPMENT

State of the sciences

The science of husbandry, hatching and rearing marine fish and shellfish has reached maturity and now is the time to turn the science into a business. We are not very good at it in Britain. Many sound scientific projects have not been picked up by industry, and the scientists have continued their work in the hope that some day it will take off, only to find that it eventually does so in another country.

The attitude of the government is, quite correctly, to make science more exploitable by using it to increase the wealth of the nation. We believe that fish farming is at this point. If we can stimulate commercial investment in the next stages, then this must be the challenge of the next few years. The formation of the British Halibut Association, a consortium of interested parties, is a step in this direction. What then are the issues to resolve?

Marketing

Crucial to the success of marine farming is a marketing policy. It is not enough to simply produce the goods in the hope that there is a market somewhere.

There is still sufficient wild stock for such species as scallops to fix the price whether they come from a farm or from the sea bed. However, farmed products do have market advantages which should be exploitable:

1 Regularity of supply, independent of weather and, to a large extent, season. The spawning times can be controlled, as has been demonstrated with turbot.
2 Uniformity of size. This could be a key advantage to the restaurant trade, which may want to put on the table a 4 oz (110 gm) portion.
3 Controlled-quality products.
4 Shellfish may be sold live, entirely free from grit and of contamination from pollution.

Fish farming in Britain is tending towards the small-scale operation or crofter who, as an individual, is not able to press forward these market advantages and may not survive in a competitive situation. There are really two choices – for the small-scale growers to band together to form marketing co-operatives, or for the bigger companies to take a share in fish farming and, in so doing, bring to bear their own marketing skills.

Cost effectiveness

The scientists' job is to solve scientific and technical problems and generally does not address the cost effectiveness of the eventual process. In marine finfish farming two areas emerge as the significant cost centres: feed, 30–35%, and labour, 20–26%.

Food conversion ratios vary but the industry depends on low-cost fresh fish or proprietary feeds. A lot of valuable research could now be done to lower food costs, and Seafish are working with the Association of Fish Meal Manufacturers and Heriot-Watt University in understanding dietary requirements and low-cost diets from fish meal. However, it will be difficult to achieve any major savings, as the quality of the product eventually depends largely on feed quality.

Labour costs are controllable either by economy of scale or by semi-automated holding systems involving feed dispensers and the mechanization of such functions as harvesting, net cleaning and processing. Regrettably there is no fish farm service industry in Britain, which depends heavily on Scandinavian imports. A lot more could be done to develop our engineering knowledge and provide a service to the industry.

Shellfish, of course, do not require food, being filter feeders. Labour costs in maintaining rafts, longlines and harvesting and grading are the dominant factor and one which would benefit from a farm service industry in the UK.

Water areas

Obtaining suitable water areas for farm sites is likely to become an increasing problem, and there is already conflict with other water users, including fishermen and yachtsmen.

The sea bed and the foreshore belong to the Crown Estate Commissioners, except where the Crown has leased certain areas to others, sometimes centuries ago. This makes the securing of growing sites a lengthy process for the prospective grower.

For any sites involving the relaying of shellfish, *bouchot* culture, the anchoring of rafts or cages, a lease has to be obtained from the Crown Estate Commissioners. The Crown Estate is naturally looking to the lease of sea bed in this way as an additional source of revenue.

There are numerous local authority constraints which restrict the potential farmer's choice of site. Objections can be made to planning consent by environmental groups or people concerned with safe navigation.

Fish farming is an industry in its early stages and should be encouraged, but there needs to be a process of harmonization between all concerned if the new industry is not to be stillborn as a result of too many punitive restrictions.

Training and extension work

The scientific work has produced the essential understanding of husbandry, nutrition, on-growing and disease for all the species. It is important that the entrepreneurial fish farmer has a thorough grounding in these matters. The industry does not enjoy the benefits of land farming in having many years of solid practical experience to draw on. The British Halibut Association will be one means of making sure that the results of this scientific work are passed on to potential growers.

Seafish are already running, with Training Agency help, a number of short courses in shellfish culture, and a new Shellfish Unit and teaching centre is under construction at Ardtoe which will provide, for a fee, all the necessary guidance on site selection, water quality monitoring, equipment, husbandry, marketing and finance.

Financial aid

Fish farming is supported by several agencies which provide grant aid or low-cost loans. This is very encouraging and is a means by which fish farming could gather momentum in the early 1990s, provided the water space can be made available. It is important, however, that the money should be used wisely and not simply allocated to any project which seems to offer potential for creating new jobs in rural areas. Fish farms have a poor cash flow in the early years and the risks are high. Any grants or loans must take these difficulties into account and make sure the farmer has adequate resources to see his enterprise through the early years.

CONCLUSIONS

There is a wealth of scientific knowledge on the cultivation of marine shellfish and certain finfish species, and the UK has established a lead in this field. It is important to capitalize on this work by innovation and turn the science into a business. Such an approach is consistent with government science policy of encouraging exploitable research.

Important work still to be pursued will aim to reduce feed and labour costs. Whilst feed must always be a significant cost, the labour costs can be contained by good engineering design and the development of a fish farm service industry, possibly run on a co-operative basis.

It cannot be assumed that because fish and shellfish can be grown they can be sold. Farmed fish have certain key market advantages which must be exploited to find the right market niche. It will not be sensible to channel farmed products into the market in such a way that they are indistinguishable from the wild stock.

Other critical areas which the Sea Fish Industry Authority are tackling are training and extension work, controlling development through financial incentives and harmonizing the growth of fish farming with environmental issues. Seafish are optimistic that marine farming has a future.

NOTE

1 A 'several order' grants exclusive permission to an individual or company to use a defined area of the sea bed for a particular purpose, e.g. whilst anyone may relay shellfish on the sea bed no one has an exclusive right to harvest the shellfish without a several order. The several order relates to Crown sea bed up to twelve miles offshore and is obtained after extended negotiation and consultation with MAFF, DAFS, the Welsh Office or the Northern Ireland Office.

5 Deep abyssal plains

Do they offer a viable option for the
disposal of large-bulk low-toxicity wastes?

Martin V. Angel

INTRODUCTION

The disposal of large-bulk low-toxicity waste material from the
ever-growing coastal conurbations throughout the world is an environmental problem that will continue to increase. The wastes can range
from sewage sludge to industrial wastes like flyash and titanium oxide,
or contaminated dredge spoils. Land-based disposal methods such as
burial, recycling and incineration all suffer from economic and/or
environmental disadvantages. At present marine disposal into shallow
water environments is one of the major methods currently employed in
the UK. Sewage and industrial wastes are discharged down outfalls or
dumped direct from vessels, and dredge spoils are used extensively for
land reclamation even when contaminated quite heavily with heavy
metals. The Royal Commission on Environmental Pollution (1985)
considered most of these options to fall within its concept of the 'Best
Practicable Environmental Option'. Environmentalists are critical of
such procedures because the criteria on which they are based are
anthropocentric and fail to take broader environmental considerations
into account. Here a novel disposal option of discharging wastes close
to the sea bed at abyssal depths is examined for its probable
environmental acceptability and is found to offer sufficient environmental advantages and safeguards to merit a full-scale trial.

POLLUTION AND CONTAMINATION

GESAMP (1982) established an important distinction between
contamination and pollution. Contamination was defined as 'the
introduction of substances into the marine environment which alters the
concentration and distribution of substances within the ocean'. Marine
pollution is:

the introduction by Man, directly or indirectly, of substances or energy into the marine environment resulting in such deleterious effects as harm to living resources, hazards to human health, hindrance to marine activities including fisheries, impairment of quality for use of sea-water and reduction of amenities.

Similar definitions have been incorporated into the United Nations Law of the Sea Convention.

Another concept introduced in GESAMP (1982) is that of environmental capacity, which was defined as the limits to the ability of the system to absorb or to buffer the influence of contamination without showing evidence of pollution. The concept must involve considerations of scale, because of deciding what are the bounds of a particular eco-system. Marine disposal policies are usually based on the concept of maximizing dispersion, so that dilution prevents deleterious effects being expressed. However, this does raise problems of monitoring the impact of the disposal, so that there is a danger of the limits of acceptability being unwittingly exceeded as the inputs are increased. The alternative approach is to operate in an accumulative regime where the impact is high but localized and readily monitored. Here, in addition to the GESAMP criteria, it is argued that any disposal regime must meet the criteria of the World Conservation Strategy in not threatening (1) the survival of any species, i.e. the maintenance of genetic diversity, (2) the continuance of global ecological processes and (3) the sustained use of living resources.

THE ABYSSAL ENVIRONMENT

The hypsographic curve for the world (Figure 5.1) shows that ocean depths of 3000–6000 m occupy just over half the earth's surface. The sea bed at these depths generally consists of basic igneous rocks originally extruded along the mid-oceanic ridge during crust formation, which are covered by pelagic sediments which increase in thickness towards the continental margins. These pelagic sediments are not usually more than a few hundred metres thick until they come within the influence of the continental margins and terrigenously derived material appears more and more abundantly. Apart from manganese nodules, which occur in rather localized concentrations, mostly in the Pacific and Indian Oceans, these oceanic sediments seem to have little, if any, potential for mineral exploitation. Their potential for disposal of radioactive wastes has been recognized for many years and has been actively studied, although not generally accepted.

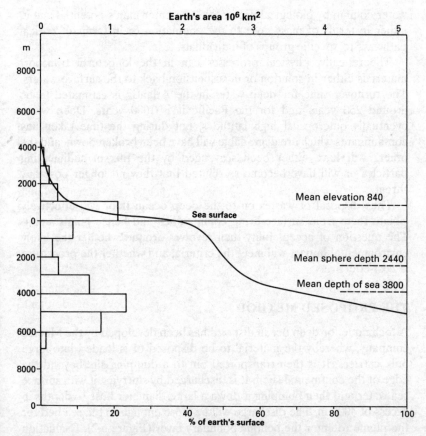

Figure 5.1 Hypsographic curve, showing the area of the earth's surface above any given elevation, and covered by ocean of any given depth. Note that over 50 per cent of the earth's surface is covered with sea 3000–6000 m deep

The living pelagic and benthic communities which inhabit the deep ocean show an exponential decrease in biomass with depth, as they become increasingly more remote from their primary sources of energy in the surface euphotic zone (Angel and Baker 1982; Lampitt *et al.* 1986). Even beneath the most productive waters the abundances of the organisms at abyssal depths are extremely low. There are very few species whose known bathymetric ranges extend throughout the total water column from the surface down into the abyss. The ranges of seasonal and diel vertical migrations are almost all encompassed within the surface 1500 m of the water column (Angel 1986). Hence there are no pathways whereby contaminants will be transported back up the water column by biological processes to re-enter man's potential ambit, either in terms of mass dose to the population or in terms of critical pathways to specific groups of individuals.

Theoretically, physical processes can in the long term transport materials either in solution or in suspension back to the surface waters. The turnover time for deep water in the Atlantic is estimated to be around 250 years and for the Pacific it is 1000 years. Deep water eventually outcrops at high latitudes, but during the time taken any contaminants which are degradable will have been broken down, and the others will have either been scavenged by the flux of sedimenting particles or will have become so diluted that they no longer pose any threat.

Thus disposal of wastes on to the deep ocean floor will effectively isolate them from surface-living communities and hence man's ambit. The question of acceptability then revolves around whether the likely environmental impact will meet the criteria, and whether the procedure is cost-effective.

THE PROPOSED METHOD

A technique for deep ocean disposal has been developed by the Maersk company, whereby the material to be disposed of is loaded into large bulk carriers. It is then transported out to a dumping site beyond the edge of the continental shelf. It is discharged by slurrying it with surface sea water and then pumping it down a large-diameter hose to depths in excess of 4000 m. The discharge will be close enough to the sea bed for the plume to enter the benthic boundary layer (Figure 5.2). Evaluation of the engineering requirements have shown that the technology exists to provide a suitable hose which can withstand the fifty-year storm in the Western Approaches with an ample safety margin. Cost assessments have not yet been finalized because of the need to select and evaluate potential disposal sites. While it will be more expensive than the present procedures, it will save on considerable capital costs which will be incurred soon in replacing barges at present used for dumping sewage sludge in the North Sea, and the renewal of outfalls. The extra cost will be offset against the improvement in the quality of inshore waters.

A statement was issued after the 1987 interministerial meeting on the North Sea to the effect that sewage sludge and discharges of partially treated material from outfalls present no environmental hazard and may even be beneficial. However, it is evident that these activities have hampered the development of a UK shellfish industry in many localities; there remain many unanswered questions about the health hazards, and

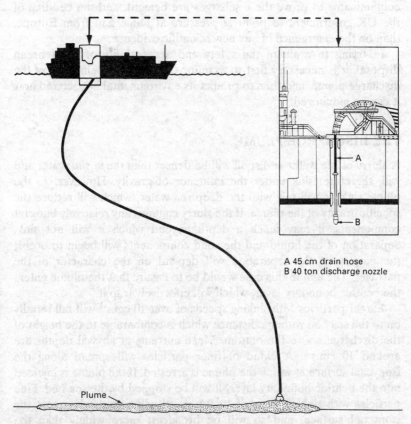

A 45 cm drain hose
B 40 ton discharge nozzle

Figure 5.2 Diagrammatic representation of the proposed disposal procedure
Courtesy Maersk

public concern is still acute (e.g. Marine Forum 1990). The assurances issued by government were based mainly on surveys in the immediate vicinity of dispersive dumping sites which were not conducted at relevant scales in either time or space to detect any changes induced. However, surveys at the accumulative site at Garroch Head in the Clyde have detailed the changes, and it is interesting to note that this site is the centre of an active local fishery, and the fish do not appear to suffer from the notorious ulcers seen in the North Sea.

In 1990 Britain announced that its policy was to halt all dumping of sewage by 1998 and that discharges down outfalls will be phased out at some unspecified date. This decision seems to be based more on the application of the precautionary principle (i.e. the onus is on the contaminator to prove the discharges are benign), and the bending of the UK government to political pressure at home and from Europe, than on the emergence of any new scientific evidence.

In trying to evaluate the safety and acceptability of deep ocean disposal, it is necessary first to examine the probable behaviour of the discharge plume, and then to predict its environmental impact and how it can be monitored.

THE DISCHARGE PLUME

A slurry of sea water and spoil will be denser than the *in situ* water and will, therefore, sink under the influence of gravity. However, as the plume sinks it will mix with the deep sea water, which will reduce the specific gravity of the plume. If the slurry contains any relatively buoyant components, it may reach a depth beyond which it will not sink. Separation of the liquid and the solid components will begin to occur; the pattern of the separation will depend on the character of the particles. The aim in this case would be to ensure that the plume enters the benthic boundary layer, which will effectively trap it.

Larger particles with sinking speeds of over 10 $cm.s^{-1}$ will fall rapidly on to the sea bed within a distance which is comparable to the height of the discharge above the bottom. Mean currents at abyssal depths are around 10 $cm.s^{-1}$. A cloud of finer particles will spread along the isopycnal surface at which the plume is arrested. If the plume is injected into the benthic boundary layer, it will be stopped by the sea bed. Fine particles with sinking rates of 10–0.1 $cm.s^{-1}$ will sink slowly across the isopycnal surface, and so will be broadcast more widely than the immediate area of the disposal. The silt/clay fractions, however, will persistently remain in suspension and may not sediment on to the sea bed for weeks or even months. The observed behaviour of the nephels described by Dickson and McCave (1986) probably provide the best naturally occurring example of the way in which this silt/clay cloud may disperse. However, sedimentation rates may be enhanced through the scavenging action of sinking 'snow' aggregates (Alldredge and Silver, 1988), or by the feeding of suspension feeders, or through flocculation. The dispersion of the silt/clay fractions is important because the greater surface area of the particles provides more chelation sites for organic molecules and also for metallic ions, so they will tend to carry the main load of contaminants.

The scavenging by the snow aggregates will be highly variable both in time and in space. They consist mainly of the products of biological

Table 5.1 Sinking speeds of particles with a specific gravity of 2.5 in still sea water with temperature and salinity characteristics typical of bottom waters in the north-eastern Atlantic

	Particle diameter (um)	Fall speed (cm.s^{-1})	Approximate time to sink 100 m
Silts	1	5.10^{-5}	6 years
	5	1.10^{-3}	3 months
	10	5.10^{-3}	20 days
Sands	50	0.1	1 day
	100	0.5	6 hours
	1000	15	10 minutes
Gravel	10 000	100	2 minutes

production in the overlying euphotic zone together with terrigenous inputs reaching the oceans via aeolian and riverine sources or from coastal erosion. Table 5.1 lists sinking rates for particles with specific gravity of 2.5 in still sea water with bottom-water characteristics of salinity and temperature. Particles with diameters less than 100 μm will obey Stokes's law, so their sinking rates will be proportional to their density difference from the surrounding sea water.

Once it reaches the sea bed, newly deposited material may be subject to resuspension, long-term burial or bioturbation. Erosion of 10 μm particles will theoretically begin to occur once currents, measured 10 m above the sea bed, have reached velocities of 20 cm.s^{-1}. Currents of even these relatively slow speeds are quite infrequent at abyssal depth, and when they do occur they are termed benthic storms. Erosion of clays, once they have been deposited, occurs less readily than expected, because they tend to be cohesive. Another factor influencing resuspension is the roughness of the sea bed; microtopography generated by bioturbation or minor geological features can cause local differences in the balance between erosion and deposition.

IMPACT ON BENTHIC AND BENTHOPELAGIC ECO-SYSTEMS

Four categories of impact are to be expected:

1 Smothering of the sedentary benthic community with an inert covering.
2 Clogging of the filter-feeding mechanisms of benthic suspension feeders.

3 Reductions in the *in situ* dissolved oxygen levels through the biological and chemical oxidation of organic material in the wastes. This will affect both the water column and the underlying sediment.
4 Toxicity of heavy metals and industrial residues in the materials being disposed of.

The characterization of the wastes will determine the relative importance of each of these impacts.

Smothering will occur with all materials, but will be restricted to the near-field of any disposal site. The effects will mimic the natural phenomenon of slumps and turbidity flows. These cause catastrophic destruction of those benthic communities which are buried. Recolonization of areas affected by such natural events has received little attention, but, whereas some slides seem to be recolonized rapidly, others off the coast of north-west Africa have not been recolonized after several centuries. Some of these flows have been shown to cover 10^3–10^4 km^2, and so extensive perturbations are known to occur naturally in abyssal plain areas, and the communities are probably adapted to cope with such large-scale disturbances. Furthermore the seasonal deposition of phytoplankton detritus at latitudes above 40° (Lampitt 1985) must also serve to pre-adapt the abyssal fauna to the problem of burial. The factor determining the rate of recolonization would seem to be the nature of the surface.

The worst-case scenario would be to assume that no recolonization ever occurs. If an arbitrary limit to destruction of the abyssal sea bed is set at 1 per cent of the total available area, then in the north-eastern basin of the Atlantic where abyssal depths extend over an area of about $1.5.10^6$ km^2, about 10^4 km^2 could be utilized for disposal. A single disposal of 10^6 m^2, about 10^4 km^2 could be utilized for disposal. A single disposal of 10^6 m^3 of slurry would have an impact over about 600 km^2 of sea bed. Thus if the areas where dumping could be undertaken were restricted, the limit suggested here would not even be approached. An important factor to bear in mind is that the zoogeographic distribution patterns of abyssal benthic communities are very uniform and do not have the medium-to-coarse scales of pattern which dominate terrestrial environments; the loss of 1 per cent of this uniform environment is unlikely to disrupt the ecology of the oceans.

Clogging of the feeding mechanisms of suspension feeders is one of the key factors in the sensitivity of tropical coral reefs to disturbance. The extent to which this may prove to be a problem in a deep oceanic environment is uncertain but is not expected to be serious. At abyssal depths such feeding predominates in two contrasting sedimentary regimes: where the sedimentation rates are extremely high, and where it is extremely low. Generally over abyssal plains in the north-east Atlantic neither regime prevails and suspension feeders seldom form a significant part of the benthic communities. Those that inhabit the high sedimentation regimes are likely to have effective self-cleaning mechanisms, which will limit the impact.

Reduction in dissolved oxygen levels in the water resulting from the chemical and biological oxygen demand of any organic material within the wastes will be of concern. Sewage sludges have a high organic content and hence a potentially high COD. If the disposal is sufficiently localized a portion of the waste will be buried deeply enough for the full demand not to be expressed. However, for this assessment it is assumed that the full draw down in dissolved oxygen will occur. Bottom water in the north-eastern Atlantic has very consistent characteristics of temperature (2.2°C) and dissolved oxygen (5.5 ml l^{-1}). It originates in the southern ocean, where very cold oxygen-rich water sinks from the surface to the depths. It flows up the western side of the southern Atlantic (the eastern side is blocked by the Walvis Ridge between the Mid-Atlantic Ridge and south-west Africa). It flows through the Mid-Atlantic Ridge via the gaps provided by the Romanche Trench on the Equator, and the Vema Fracture Zone at 10°N, and so enters the eastern side. It flows northwards to enter the north-eastern basin through Discovery Gap (37°20'N, 15°34'W). The changes in the water's characteristics during its passage from the southern ocean to high northern latitudes in the various oceans are illustrated in Figure 5.3.

The flow through Discovery Gap has been estimated at about one sverdrup (i.e. 10^6 m^{-3} s^{-1}). This estimate of inflow can be used to set a provisional limit to the quantities of oxidizable material that can be introduced into the deep water. Setting an arbitrary limit to the maximum permissible draw down in the mean dissolved oxygen at 0.5 ml l^{-1}, the renewal rate of water through Discovery Gap will be sufficient to oxidize annually 150.10^6 tonnes of the sort of sewage sludge which is being dumped at present in the Thames estuary.

This estimate can be set in the context of the most recent published data on how much of its annual production of sewage sludge the UK dumps around the coasts. It amounts to 10.10^6 tonnes, which is 30 per cent of the total production. This does not include local discharges down sewage outfalls which dispose of even more partially treated sewage (Royal Commission on Environmental Pollution 1985).

If all this sludge was disposed of at a single site 60 km in diameter where the benthic boundary layer is about 50 m in thickness, the benthic boundary layer would need to turn over about three times in a year to

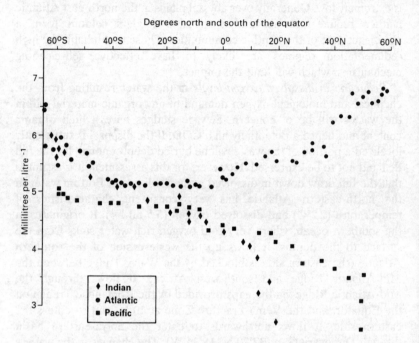

Figure 5.3 Latitudinal distribution of dissolved oxygen (ml. l^{-1}) in the bottom waters of the western regions of the three main oceans (modified from Mantyla and Reid 1983). Note that in the eastern Atlantic the concentration of dissolved oxygen does not increase north of 35° N

maintain the suggested level of dissolved oxygen within the area of influence of the disposal site. Another effect that must also be borne in mind is the effect of reductions in dissolved oxygen on sediment chemistry, and this will be discussed below.

Metal toxicity will be of concern, particularly for dredge spoils and industrial wastes. Disposal of these wastes in shallow water depends on the dispersive effects of inshore currents to keep the levels from rising too high. Even so, the accumulation of heavy metals in estuarine environments is causing concern, especially where the estuaries are used heavily for fisheries and recreation (North Sea Forum 1987). In the oceans generally, metals and other naturally occurring compounds can be classified into three main categories, depending on the degree to which biological processes influence their profile. Bio-limited compounds are tightly regulated by biological processes to such an extent that in stable water columns their concentrations can be reduced below levels of detectability. Bio-intermediate substances are those which are affected by biological processes but physio-chemical processes largely determine their concentration profiles. Bio-unlimited substances are totally unaffected by the biology. Generally the concentration profiles of tracers in the oceans are determined by the long-term balance between inputs and sinks, and these may vary quite considerably between the main ocean basin. Deep disposal will create a localized sea bed input of metal contaminants in the waste being disposed.

Heavy metal levels in sewage sludge in the UK have been declining as the water authorities have been increasingly successful in keeping domestic and industrial waste streams separate. The sludge dumped in the Thames estuary at present contains about 1 ppm of heavy metals. The concentration ranges of various contaminants in dredge spoil from the Mersey in terms of $\mu g\, g^{-1}$ dry weight are: Ag 0.34–1.8, As 10.3–70.9, Cd 0.2–3.9, Cr 37–142, Cu 31–144, Fe3 1.7–3.9, Hg 0.4–6.2, Mn 633–2183, Ni 17.7–44, Pb 68–205, Sn 2.4–12.9, Zn 217–627. High though some of these figures are, some estuaries in the south-west of Britain contain even higher levels of some of these metals (e.g. As 2520, Cu 2540 and Zn 3515 in Restronguet Creek, Langston 1986). However, whether a metal is taken up by an organism with advantageous or deleterious effect is not entirely dependent upon the concentration of the metal in the environment, but also upon its availability. Metals may be taken up by direct absorption from solution, by uptake of particulate matter or via their food. In terrestrial food webs bio-magnification occurs, whereby top predators accumulate toxic levels of contaminants via their food. Despite our general lack of detailed knowledge about deep oceanic food webs, there seems little reason to suspect that similar problems will occur in the deep ocean.

The effects on sediment chemistry must also be considered. In their normal state the upper layers of oceanic sediments remain well oxygenated because of the delicate balance maintained between the utilization of the oxygen through respiration and chemical oxidation and the resupply by diffusion. The system can cope only with small enrichments. Moreover the dumped material will isolate the sediments from the overlying oxygenated waters, even if it has no oxygen demand itself. The subsequent decrease in the redox potential within the superficial layers of the sediment will result in the mobilization of ions such as iron, manganese, nickel, uranium, vanadium and zinc, possibly making them biologically more available. Such insulation of sediments from the overlying water occurs beneath turbidite deposits, and the subsequent sequence of events is known in outline (Jarvis and Higgs

1987). While these turbidity flows have not been shown to generate environmental perturbations through the changes they produce in the sediment chemistry, it would be wise to avoid areas of previous dumping activity, such as old disposal of radioactive wastes.

CRITERIA OF ACCEPTABILITY

Any activity has an environmental cost, and the decision as to which is the Best Practical Environmental Option (BPEO) has to be a pragmatic judgement balancing financial against environmental costs. The recent adoption of the 'precautionary principle', putting the onus of proof on the polluter, will make waste disposal an impossible activity, because any waste disposal causes environmental damage. Whereas the generation of many types of waste can be curtailed or even stopped, some simply cannot. It needs to be recognized that not all eco-systems are of equal value, or vulnerability, or fragility. Any innovative procedure needs to be thoroughly scrutinized within a global context especially if it may violate existing international conventions and protocols. It is quite clear that European shelf seas, and the North Sea in particular, are under considerable environmental pressure as a result of the multiplicity of uses to which they are put (e.g. Smith and Lalwani 1984).

Although these seas are robust systems they are showing signs of incipient problems, e.g. increases in eutrophication and in the incidence of 'red tides' (North Sea Forum 1987). In the North Sea many of the problems tend to be concentrated outside British waters in the Wadden See and the Kattegat, but recent modelling data suggest that Britain may well contribute to the problems being experienced in these areas (Hainbucher *et al.* 1987) (Figure 5.4). Causal relationships are extremely difficult to establish, so the evidence is treated as equivocal. Nor is it considered serious enough for enough resources to be provided to confirm or deny the need for the sort of active cleaning up that was taking place in British rivers. Thus in the UK until 1990 there was little perception of a need for alternative waste disposal options. Nevertheless, from what is known about the deep oceans, there seems to be a *prima facie* case that they do offer disposal options that will isolate the wastes effectively, and that such disposal will threaten neither existing or foreseeable oceanic resources, nor violate the main tenets of the World Conservation Strategy.

The technology now exists whereby adequate monitoring of such disposal sites could be achieved. All the relevant parameters suggested by GESAMP (1980) can be assessed, although the cost of the monitoring would be a small but significant element in the overall cost

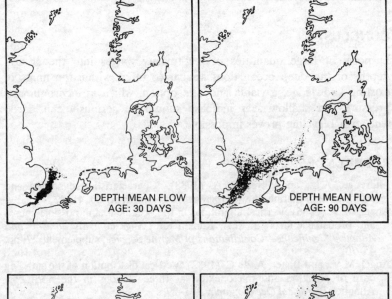

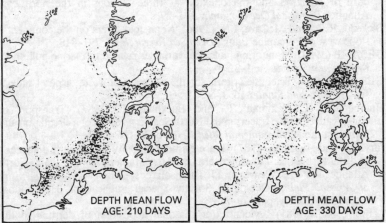

Figure 5.4 Distribution of passive tracers in the North Sea from a source near the Thames estuary (from the model output of Hainbucher *et al.* 1987)

of the procedure. The monitoring would also be important in reassuring public opinion that the disposal is indeed environmentally safe. However, any large-scale use of the oceans in this manner would need an international regulatory body to ensure that any disposal 'upcurrent' in the deep water – for example, off South America – did not depress the dissolved oxygen levels to such an extent that disposal in the North Atlantic would have to be curtailed to keep within any guidelines eventually proposed.

CONCLUSION

Disposal of large quantities of low-toxicity wastes into the abyssal depths of the deep ocean does appear to offer, within the limits of existing knowledge, a viable and safe option, which as environmental pressures on shallow seas increase should be seriously considered whenever recycling proves impossible.

REFERENCES

Alldredge, A.L., and Silver, M.A. (1988), 'Characteristics, dynamics and significance of marine snow', *Progress in Oceanography* 20: 41–82.

Angel, M.V. (1986) 'Vertical migrations in the oceanic realm: possible causes and probable effects', in M.A. Rankin (ed.) *Migration: Mechanisms and Adaptive Significance, Contributions in Marine Science*, supplement 27, pp. 45–70.

Angel, M.V., and Baker, A. de C. (1982) 'Vertical distribution of the standing crop of plankton and micronekton at three stations in the north-east Atlantic', *Biological Oceanography* 2: 1–29.

Dickson, R.R., and McCave, I.N. (1986) 'Nepheloid layers on the continental slope west of Porcupine Bank', *Deep Sea Research* 33: 791–818.

GESAMP (1980) 'Monitoring biological variables related to marine pollution', *Reports and Studies* 11: 1–22.

GESAMP (1982) 'Scientific criteria for the selection of waste disposal sites at sea', *Reports and Studies* 16: 1–60.

Hainbucher, H., Pohlman, T., and Backhaus, J. (1987) 'Transport of conservative passive tracers in the North Sea: first results of a circulation and transport model', *Continental Shelf Research* 7: 116–80.

Jarvis, I., and Higgs, N. (1987) 'Trace element mobility during early diagenesis in distal turbidities: Late Quarternary of the Madiera Abyssal Plain, North Atlantic', in P.P.E. Weaver and J. Thomson (eds) *Geology and Geochemistry of Abyssal Plains*, Geological Society Special Publications 31, London: Geological Society, pp. 179–213.

Lampitt, R.S. (1985) 'Evidence for seasonal deposition of detritus to the deep-sea floor and its subsequent resuspension', *Deep Sea Research* 32: 885–97.

Lampitt, R.S., Billett, D.S.M. and Rice, A.L. (1986) 'Biomass of the invertebrate megabenthos from 500 to 4100 m in the northeast Atlantic Ocean', *Marine Biology* 93: 69–81.

Langston, W.J. (1986) 'Metals in sediments and benthic organisms in the Mersey estuary', *Estuarine, Coastal and Shelf Science* 23: 239–61.

Mantyla, A.W., and Reid, J.W. (1983) 'Abyssal characteristics of the world's ocean waters', *Deep Sea Research* 30A: 805–33.

Marine Forum for Environmental Issues (1990) *North Sea Report*.

North Sea Forum (1987) *Report, March 1987*, London: North Sea Forum.

Royal Commission on Environmental Pollution (1985) *Eleventh Report*. *Managing Waste: the Duty of Care*, London: HMSO.

Smith, H.D., and Lalwani, C.S. (1984) *The North Sea: Sea Use Management and Planning*, report by North Sea Research Unit, Centre for Marine Law and Policy, UWIST, Cardiff.

6 The law of marine nature conservation in the United Kingdom

John Gibson

INTRODUCTION

The first statutory marine nature reserve in Great Britain was established around the island of Lundy in the Bristol Channel in November 1986, five years after the passage of the Wildlife and Countryside Act 1981, which provided the necessary legislative powers. Plans for a second reserve at Skomer and Marloes in south-west Wales are approaching fulfilment, but have been bedevilled by procedural problems. The purpose of this chapter is to examine the role played by the law in the protracted process of marine nature conservation, and to consider the wider implications for the use of legal machinery in the management of the coastal zone.

LEGISLATIVE BACKGROUND

The ability to declare national and local nature reserves on land has existed for nearly four decades under the National Parks and Access to the Countryside Act 1949.[1] The legal requirements for land-based conservation, however, differ substantially from the demands of the marine environment. The principal protection for a terrestrial nature reserve is achieved through the exercise of private property rights, whereby the Nature Conservancy Council or a local authority either acquires title to the land or enters into an agreement with the landowner. Ownership ensures that a site is not used in a way that conflicts with the objectives of conservation. The additional sanction of the criminal law through the imposition of byelaws is necessary only to control the conduct of the public in so far as they are able to have access to a reserve. At sea, on the other hand, only the soil of the foreshore or sea bed is capable of ownership, and the water is an unappropriated element subject to general rights of navigation and fishery by the public.

The wide range of commercial, industrial and recreational activities that take place within inshore waters are based upon freedom of access to the sea; they in turn attract the jurisdiction of a variety of governmental bodies, which are empowered to control or promote individual uses of the coastal zone. In contrast, the function of a marine nature reserve is to manage an area of the sea itself rather than a particular activity, and it thus faces a potential conflict with these existing rights and powers.

The possibility of nature reserves in the shallow seas was suggested as long ago as 1968 by the former Nature Conservancy, but was initially rejected by the Natural Environment Research Council (NERC) for lack of evidence that a need existed. However, a report in 1973 by a NERC Working Party on Marine Wildlife Conservation (NERC 1973) led to joint research with the Nature Conservancy Council, and culminated in a further report (NCC/NERC 1979) which recommended that consideration be given to obtaining legislation for the protection of areas below low water mark. In a diplomatic attempt to pre-empt opposition from government departments and statutory bodies already involved in coastal administration, a working party on marine nature reserves was formed, including representatives from the Departments of the Environment, Energy, Transport, Trade and Industry, the Foreign and Commonwealth Office, the Ministries of Defence and Agriculture, Fisheries and Food, the Crown Estate Commissioners, the Health and Safety Executive, the Scottish Development Department, the Department of Agriculture and Fisheries for Scotland and the Welsh Office. The inevitable consequence of such well intentioned consultation was to attract theoretical objections based upon speculative fears about the way in which conservation might be used. It is, of course, hardly surprising that civil servants should regard it as their duty to defend the section of the public interest represented by the department to which they belong. Under such circumstances, the need to establish priorities and subordinate one advantage to another is most unlikely to be achieved, and the price for the withdrawal of opposition becomes a guarantee of immunity from the consequences of reform.

THE WILDLIFE AND COUNTRYSIDE ACT 1981

When the Wildlife and Countryside Bill was introduced in the House of Lords in November 1980 it contained no provision for marine nature reserves. Instead the Department of the Environment (1981) issued a public consultation paper, inviting responses by August 1981, and effectively ensuring that the opportunity for legislation would be missed. However, two peers, Lord Mowbray and Stourton and Lord

Craighton, sought to put pressure on the government by successfully introducing three separate amendments,[2] proposing alternative forms of marine reserve. Faced with a demonstration of political will, the government responded by producing its own more restrictive scheme in standing committee;[3] this was subsequently replaced by a revised version in the House of Commons,[4] and, after further amendment in the House of Lords,[5] was enacted as sections 36 and 37 of the Wildlife and Countryside Act 1981.

The resultant procedure for the declaration of marine nature reserves is exceptional, and contrasts markedly with that applicable to reserves on land. A major difference is the indirect method by which the Nature Conservancy Council may acquire the right to manage an area of sea. The statutory power to make a designation order is entrusted not to the council itself but to the Secretary of State for the Environment, Scotland or Wales;[6] it was considered more appropriate that the legislation should be implemented by an accountable government minister rather than a statutory body, in view of the greater public impact of a marine reserve. Nevertheless, the process must still be initiated by the Nature Conservancy Council, which makes an application to the Secretary of State. That application must be accompanied by draft byelaws proposed by the council for the protection of the reserve.[7] The power to make byelaws for the purpose is a general one, but it specifically includes the prohibition or restriction of the following activities:

1 The entry into, or movement within, the reserve of persons and vessels.
2 The killing, taking, destruction, molestation or disturbance of animals or plants of any description in the reserve, or the doing of anything which will interfere with the sea bed or damage or disturb any object in the reserve.
3 The despositing of rubbish in the reserve.[8]

The apparent scope of such controls is, however, deceptive, because they are severely qualified by other provisions. First, neither a marine nature reserve nor its byelaws may interfere with the functions of eleven 'relevant authorities',[9] identified as local authorities, water authorities, internal drainage boards, navigation authorities, harbour authorities, pilotage authorities, lighthouse authorities, conservancy authorities, river purification boards, district salmon fishery boards and local sea fisheries committees.[10] This list effectively grants immunity to every local body that exercises jurisdiction in the coastal zone. Second, the same privileged status is accorded to any functions conferred by Act of

Parliament, and thus extends equally to central government departments.[11] Such limitations demonstrate how the political answer to the problem of overlapping administrations is to offer a negative guarantee of non-interference rather than to make a positive attempt at reconciliation. Likewise, all private rights take precedence within a marine reserve;[12] these will include rights of property in the foreshore and, in some places, private fisheries created by the Crown before 1189 or, in the case of shellfish rights, granted subsequently by legislation. The Crown Estate Commissioners are, of course, also the predominant owners of the foreshore and territorial sea bed, and no reserve may be established on their property without consent.

The breadth of interests that are permitted to prevail over a marine nature reserve is a significant indicator of the cautious legislative policy adopted in the coastal zone. Indeed, in some respects the limitations are more theoretical than real, since it is difficult to envisage how conflict could arise with some of the 'relevant authorities', who seem to have been included merely for the sake of completeness. On the other hand, concern about the relationship with the governmental bodies responsible for industrial fishing, which was a major source of contention during the parliamentary debates, has proved to be fully justified in practice.

A further feature that emphasizes the placatory approach of the Wildlife and Countryside Act is the remarkable attempt to utilize the existing regulatory functions of 'relevant authorities' instead of concentrating control in the hands of the Nature Conservancy Council. The intention is that such authorities should be invited to make byelaws under their own enabling legislation for the benefit of a marine nature reserve; these byelaws should accompany an application for the designation of a reserve, thereby minimizing the ostensible need for restrictions imposed by the council, and compensating for the weakness of its powers. This is, of course, an essentially diplomatic device designed to enlist the support of other administrative bodies while simultaneously guaranteeing their independence. Unfortunately, it also presupposes a degree of mutual co-operation that experience has shown to be unrealistic.

There is evidence too of the supremacy of commercial considerations in the exclusion from navigational constraints of all vessels except 'pleasure boats'.[13] Because of the discriminatory effect of this provision, lobbying by yachtsmen successfully confined the remaining restrictions to part rather than the whole of a reserve and to particular times of year. Not only, therefore, is the passage of vessels subject to minimal regulation, but the legislation also omits to provide any definition of the 'pleasure boats' to which it does apply. It was mistakenly supposed in the House of Commons that such craft were already defined in the Public Health Acts.[14] Even if this assumption were correct, it would have been of little assistance, since the technical meaning of a word in one statute is relevant only to the context in which it is used, and cannot safely be imported elsewhere. There are, in fact, several statutory definitions of 'pleasure craft' in customs and harbour legislation, which concentrate variously on the size of a vessel, the number of persons on board or the purpose for which it is used.[15] Such tests would not provide an unequivocal answer, for example, to the status of a commercial boat hired to carry passengers into a marine reserve for recreational purposes. It is precisely that kind of borderline case that could give rise to unnecessary litigation.

Another anomaly, which has now been cured by the passage of time, is the inability of byelaws to prohibit the discharge of pollutants from vessels in a marine reserve. This was again due to an erroneous assumption[16] that the matter was already covered by the legislation on dumping at sea, which actually relates to the deposit of substances or articles. Fortunately, the omission was counteracted by the eventual implementation of Part II of the Control of Pollution Act 1974 in 1984–5, which provides an alternative remedy.[17]

IMPLEMENTATION OF THE LEGISLATION

The passage of the Wildlife and Countryside Act, even in a flawed condition, is of course something of an achievement. Nevertheless, the presence on the statute book of enabling provisions for the designation of marine nature reserves is an empty victory unless the powers are capable of successful implementation. The conspicuous delay in the creation of reserves is at least partly attributable to the inherent weakness of the statutory powers, and is a direct consequence of the compromises that were made in order to appease opponents of the legislation. The two-tier system whereby the Nature Conservancy Council must persuade the Secretary of State to endorse its proposals inevitably prolongs the procedure. In principle, it is the duty of the Secretary of State himself to consult interested parties, publicize draft orders and consider representations before designating a reserve.[18] In practice, he will not entertain an application unless the Nature Conservancy Council has first issued its own consultation paper and conducted exhaustive negotiations with objectors. Legal responsibility is thus separated from practical involvement, and the role of the Secretary of State becomes that of an arbiter who naturally prefers consensus to confrontation. The withdrawal of opposition becomes an objective rivalling the needs of conservation, and in such circumstances the agreed solution may easily be an agreement to do nothing.

Lundy Marine Nature Reserve

Public relations are a crucial element in the success of a proposal. It was for this reason that the first location suggested for a marine nature reserve, the Scilly Isles, was abandoned in the face of local hostility to interference in the affairs of that community. The next attempt, at Lundy Island, did indeed result in the establishment of a reserve,[19] but considerable difficulty was encountered before that outcome was eventually achieved. Lundy is a remote island eleven miles off the Devon coast, which is managed by the Landmark Trust under a lease from the National Trust; it was already the site of a voluntary marine nature reserve, and might thus have been expected to present few conflicts of interest. The area selected for the statutory reserve is the sea and its bed between parallels of latitude and longitude approximately 1500 m from the shore. This is substantially less than the permissible maximum of three nautical miles from the baselines of the territorial sea. At present the waters considered suitable for marine nature reserves are all in close proximity to land, and, when the territorial sea was extended to twelve miles on 1 October 1987, the opportunity to increase the scope of the Wildlife and Countryside Act was rejected, although the possibility remains of future enlargement by order-in-council.[20]

The landward limit of the Lundy reserve is more problematic. A marine nature reserve can be created only in relation to 'land covered (continuously or intermittently) by tidal waters or parts of the sea'.[21] It was the intention of Parliament that the legislation should not be confined to sub-tidal regions, but should also apply to the foreshore between high and low water marks. In legal usage, the boundaries of the foreshore are calculated by reference to medium tides, although mean spring tides are also used for some jurisdictional purposes. However, the Nature Conservancy Council has adopted the position of highest astronomical tides to delimit both the Lundy and the proposed Skomer reserves. This criterion represents the highest level that can be predicted to occur under average meteorological conditions and any combination of astronomical factors. Since it follows a theoretical cycle of approximately 18.6 years, it is arguably too infrequent an event to justify the description of 'intermittent'. Moreover, the scientific reason for including this inner margin is not because it is intertidal but in order to incorporate the 'splash zone' where fauna and flora are affected by salt water spray above the ordinary high water mark. It is precisely because the area is not regularly submerged by the tide that the Nature Conservancy Council wishes to protect it. The possibility arises, therefore, that the statute may not support the jurisdiction that is claimed, and a provision that is supposed to facilitate conservation instead becomes an impediment to its exercise. This problem is typical of the distortions produced by a system in which proposals for reform must be translated by parliamentary draftsmen into the legal language that is a prerequisite of their implementation.

The byelaws made by the Nature Conservancy Council for Lundy are very basic, and simply repeat the prohibitions on interference with animals, plants, objects or the sea bed, and on the deposit of rubbish, which are contained in section 36 of the Wildlife and Countryside Act.

Significantly, no restrictions are imposed on navigation by pleasure boats, and the byelaws even stress that all rights of passage are unaffected. There is also a clause preserving rights of fishery, whether public or private, which consequently excludes fish from the scope of the byelaw against killing or taking animals. The control of fishing, which is perhaps the major threat to a marine nature reserve, is thus beyond the legal capacity of the Nature Conservancy Council responsible for the management of the site. Instead the role is entrusted to the local sea fisheries committee, which is a 'relevant authority' with jurisdiction up to three miles from the coast, and is entitled to make byelaws under the Sea Fisheries Regulation Act 1966.[22] Initially, however, the sea fisheries committees refused to fulfil the optimistic expectations of the Wildlife and Countryside Act, and declined to exercise their powers for the purpose of nature conservation. They argued that they were statutorily confined to considering the long-term welfare of the commercial fishing industry, and would be breaking the law if they acted for an ulterior purpose. The fact that public authorities are creations of statute, and may do only what is expressly or impliedly permitted by legislation, is a serious impediment to co-operation, unless Parliament provides for greater flexibility in its enactments.

It might be supposed that, if a 'relevant authority' disclaims jurisdiction, the matter should revert to the residual competence of the Nature Conservancy Council. However, the council takes the more cautious view that it cannot regulate any activity which is potentially controllable by another body, irrespective of the purpose for which such jurisdiction exists. This seems unduly restrictive, since the prohibition in the Wildlife and Countryside Act[23] on interference with the functions of 'relevant authorities' may only mean that the Nature Conservancy Council must not frustrate the exercise of their powers; the introduction of negative controls over the public by one authority does not necessarily preclude simultaneous regulation by another.

The *impasse* created by the attitude of the sea fisheries committees attracted criticism from the House of Commons Environment Committee (1984–5: I, paras 70–2), which recommended in January 1985 that the Department of the Environment and the Ministry of Agriculture, Fisheries and Food should take urgent action to break the deadlock in negotiations, by legislation if necessary. In fact the initiative was taken in a private member's Bill rather than by the government. The Wildlife and Countryside (Amendment) Bill, introduced by Dr David Clark, contained a clause[24] that would have removed the power of veto from 'relevant authorities', and substituted instead a requirement of ministerial approval. This amendment, however, was defeated in standing committee,[25] where the government argued that the consultative process had not been entirely exhausted, and undertook to make a concerted effort to reconcile the protagonists. Official strategy was arguably vindicated, in as much as the sea fisheries committees were eventually persuaded to co-operate in the establishment of marine nature reserves. Nevertheless, it was nearly two years before the first reserve was finally declared at Lundy in November 1986. The contribution of the Devon Sea Fisheries Committee consisted of three byelaws. One prohibits spear fishing throughout the area; another prevents trawling and netting, but, in order to satisfy local fishermen, it is limited to part of the reserve; the third outlaws potting in a single small section.

A disappointing feature of the designation process is the relative obscurity of the ministerial order that defines a marine nature reserve and the byelaws to which it is subject. It is surely proper that legislation of national importance should be issued in a permanent and accessible form. The obvious medium for this purpose is the statutory instrument, which is printed by Her Majesty's Stationery Office, numbered in an annual series, and catalogued for future reference; it is also the method by which areas of special protection for birds are created under the Wildlife and Countryside Act.[26] Yet the same Act merely requires that the existence of a marine nature reserve order should be advertised in the *London* or *Edinburgh Gazette* and at least one local newspaper.[27] Unfortunately, public awareness of the Lundy (Bristol Channel) Marine Nature Reserve Order 1986 can hardly have been aided by its inclusion in the *London Gazette* under the heading of the wrong government department.[28] The form of publication is not prescribed, and the actual order is simply a photocopied typescript; it is obtainable only from the Bideford office of the Torridge District Council, an authority that has a purely geographical connection with the reserve. Ironically, a copy has also been officially deposited at the Tavern on Lundy itself, although its utility may be questioned, since a visitor must first pass through the marine nature reserve before he can gain access to the byelaws that should already have been observed. This kind of *ad hoc* procedure is more commonly used for orders that either relate to private property or are made by local authorities rather than central government; it is far less appropriate for the establishment of a marine nature reserve in the public domain by a Secretary of State.

Proposed Skomer Marine Nature Reserve

Lundy remains as yet the only example of a statutory marine nature reserve in Great Britain, and its belated success may be attributed in part to the modesty of the enterprise. Certainly, the problems that have beset the concurrent scheme for a reserve at Skomer result directly from their more ambitious character. The Nature Conservancy Council issued a consultation paper in July 1986, and submitted a formal application to the Secretary of State for Wales in the following December. The area scheduled for protection extends about 500 m from the Marloes peninsula and around Skomer Island, and includes the waters of the intervening sound. A major innovation in the original proposal was the identification of four 'special areas' that would be subject to additional restrictions by the Nature Conservancy Council. One of these areas consisted of five sections up to 100 m from the coast, from which it was intended that persons and pleasure boats should be excluded between March and July, unless licensed by the council; the objective was to prevent disturbance of razorbills and guillemots nesting on the cliffs during their breeding season. This plan, however, was vehemently opposed both by representatives of amateur divers, who considered that they would be unreasonably victimized, and by some conservationists, who feared that the legislation was being misapplied at the expense of public goodwill. Not even the prospect of access under licence was sufficient to divert criticism from a matter perceived as setting a potential precedent.

The first objection was that the Wildlife and Countryside Act does not authorize the use of marine nature reserves for the protection of birds. A reserve may be established for the purpose of either conserving marine flora and fauna or geological or physiographical features of special interest in the area, or providing special opportunities for study and research into such matters.[29] The argument, therefore, is that birds do not qualify as marine fauna. There is no definition of 'marine' in the

Act, and, since it is not a technical legal term, it must be given its ordinary commonsense meaning. The choice lies between confining the description to species that actually live in salt water and interpreting it more broadly to embrace those that are merely dependent on or associated with the sea. As a matter of language the expression 'sea bird' obviously implies marine connotations; an analogous usage occurs in the Convention on the Conservation of Antarctic Marine Living Resources 1980,[30] where the phrase 'marine living resources' expressly includes birds. A marine nature reserve is statutorily composed not only of water, but also of land, which at common law incorporates the superjacent air space; it thus seems reasonable that creatures which inhabit the air within a reserve should be capable of receiving its protection. There are statements in the parliamentary debates[31] indicating an assumption that birds will benefit from marine reserves; indeed, one purpose of the legislation was to enable the United Kingdom to fulfil its obligations under the Bonn Convention on the Conservation of Migratory Species of Wild Animals 1979[32] and the Berne Convention on the Conservation of European Wildlife and Natural Habitats 1979,[33] which both apply to birds. In a sense, however, any attempt to seek the intended significance of 'marine' in the Wildlife and Countryside Act is an artificial exercise; the word has simply been added to terminology previously used in the National Parks and Access to the Countryside Act 1949 to[34] describe the objectives of terrestrial nature reserves. The derivation methods of parliamentary draftsmanship suggest that the word is really nothing more than a label distinguishing the identity of two kinds of nature reserve, and serves no deliberate scientific purpose; it is yet another example of the indirect way in which the legal process can confuse the implementation of political design.

A more serious criticism is that the nesting sites of the sea birds are located outside the proposed boundaries of the Skomer reserve, whereas the legislation is supposed to be used for the protection of creatures inside the designated area. Unless cliffs overhang a marine nature reserve or rise vertically above its edges, they must be situated beyond the line of highest astronomical tides. The source of disturbance by the public would, of course, occur within that jurisdictional limit, but it is arguably unlawful to impose internal restrictions for external purposes. Perhaps, since the birds would feed in the reserve, the proposal could be supported as a safeguard for its ecology, which might be affected by the disruption of their breeding cycle. Such justifications, however, conceal the real reason why it is found necessary to apply the law in this way. The cliffs are, in fact, included in the land-based national nature reserve on Skomer, where the Nature Conservancy Council has alternative powers to make byelaws as far seaward as the mean low water mark.[35] Thus a combination of two overlapping statutory procedures allows the establishment of nature reserves throughout both dry land and coastal water. Yet this dual system inevitably fails to provide for the control within one area of activities that have a detrimental impact on the other. The problem is a typical consequence of the attempt to legislate separately for land and sea, and to introduce marine equivalents of terrestrial institutions, rather than recognize the coastal zone as a single entity requiring comprehensive treatment.

Similar obstacles have impeded the desire to create another five-part exclusion zone incorporating the haul out and pupping sites used by grey seals from March to July. In addition, the Nature Conservancy Council wished to protect a further small area, 100 m square, containing a bed of *zostera marina*, which might be damaged by vessels anchoring between February and November. In the curious belief that no power existed simply to proscribe the use of anchors, it was decided to seek the more draconian solution of prohibiting entry by pleasure boats; this seems not only excessive but also unnecessary, since anchoring is included in the legal concept of navigation, and must therefore be equally susceptible to independent control. Not surprisingly, there was again serious opposition to the restriction from groups representing diving interests. Indeed, the only uncontroversial proposal involved the imposition of a speed limit of eight knots on pleasure boats within 100 m of the shore. In the face of concerted hostility to the other 'special areas', the Secretary of State for Wales was unwilling to confirm the draft byelaws in their original form. Instead the Nature Conservancy Council was obliged to negotiate a compromise with its opponents, whereby the three exclusion zones would have no statutory force, but would be subject to a voluntary code of guidance; mooring buoys would be provided in the vicinity of the *zostera* bed, and boats would be warned of the dangers of anchoring. If this experiment proved unsuccessful the council would be able to reapply for new byelaws after three years.

Thus a statutory marine nature reserve may become little more than advisory, and legislative powers that were won with difficulty may remain unused because the persons at whom they are directed understandably find them unwelcome. While it is, of course, important that laws creating criminal offences should command general support, it has never been a condition of their existence that they should attract total agreement; indeed, in such ideal circumstances, there would be no need for law at all. Admittedly, the mere fact of statutory designation should serve to emphasize the importance of a reserve and encourage public co-operation, but it is ironic that so much effort should be expended to give notional legal status to a site at Skomer that is already run as a voluntary marine nature reserve by the West Wales Naturalists' Trust.

The only other draft byelaws promoted by the Nature Conservancy Council itself embody the basic prohibitions on interference with animals, plants, objects or the sea bed, and on the deposit of rubbish, that have already been introduced at Lundy. None of these affects fishing, but the South Wales Sea Fisheries Committee agreed to make byelaws preventing the use of dredges and beam trawls, and outlawing spearfishing and the taking of fish by divers. However, the diving organizations that had already succeeded in defeating the exclusion zones now contended that divers would be unfairly penalized by a byelaw aimed specifically at their activities. This argument illustrates the way in which a cautious approach to legislation may easily be counter-productive, since selective restrictions are more vulnerable to criticism on the basis that they are discriminatory. In a vain attempt to satisfy objectors, the byelaws was redrafted in the form of an impersonal ban on all types of sea fishing with the exception of potting and angling, but the divers remained implacably opposed to a measure that would still curtail their conduct. The Sea Fisheries Committee's byelaws also requires ministerial confirmation,[36] and, in November 1988, it was submitted to the Secretary of State for Wales, together with a reapplication for the Skomer marine nature reserve, on the ground that further negotiation would be fruitless. A decision is awaited, but the outcome has inevitably been delayed.

CONCLUSION

Preparations for other marine nature reserves are even less advanced. In September 1988 a consultation paper (NCC 1988) was issued in relation to the Menai Strait in Wales, and should be followed by another concerning Bardsey Island when the uncertainties of Skomer have been resolved. In Scotland progress on a proposal for Loch Sween is taking precedence over plans for St Abb's Head, while attention in England is now focused on the Lindisfarne area. These few locations constitute a tiny fraction of British coastal waters, and their protection will produce a relatively small impact on the environment. Yet marine nature reserves represent the only available weapon in the legal armoury of conservation measures. In contrast, nature reserves on land are supplemented by more numerous sites of special scientific interest (SSSIs), where damaging operations can be pre-empted or delayed.[37] Because the procedure for designating SSSIs involves local planning authorities,[38] whose jurisdiction does not extend beyond low water mark, it is assumed that the system cannot be applied to the sea. Nevertheless, a modest attempt in the Wildlife and Countryside (Amendment) Bill to overcome this dilemma and facilitate the notification of underwater sites was defeated by the government in 1985.[39]

The history of marine nature reserves in the United Kingdom provides a revealing illustration of the problems encountered in legislating for the coastal zone. In particular, it demonstrates the legal interdependence of land and sea, and the necessity for integrated reforms; it also shows how the diversity of public and private interests involved in inshore waters can undermine both the achievement and the implementation of statutory powers. The British approach to marine conservation has proved the weakness of over-reliance on consensus and minimal interference with the *status quo*. The coastal zone is a complex and controversial environment in which it is the duty of government to determine and enforce priorities. In the final analysis, only law can offer an effective means of translating policy into reality; but law is counterproductive if misused, and the Wildlife and Countryside Act affords a salutary example of the way in which legislation may also impede administrative action. The experience of marine nature reserves is a microcosm of the difficulties of coastal zone management; it is to be hoped that the lessons will be learned for the future.

NOTES

1 Part III
2 H. L. Deb., vol. 416, cols 1176–86 (3 February 1981); H.L. Deb., vol. 417, cols 593–601 (17 February 1981); H.L. Deb., vol. 418, cols 432–4 (12 March 1981).
3 H. C. Deb., Standing Committee D, cols 562–626 (9 June 1981).
4 H. C. Deb., vol. 8, cols 924–37 (13 July 1981); H.C. Deb., vol. 9, cols 1119–201 (30 July 1981).
5 H. L. Deb., vol. 424, cols 522–32 (15 October 1981).
6 Wildlife and Countryside Act 1981, s. 36(1).
7 ibid., s. 36(2).
8 ibid., s. 37(2).
9 ibid., s. 36(6).
10 ibid., s. 36(7).
11 ibid., s. 36(6).
12 ibid., s. 36(6).
13 ibid., s. 37(3).
14 H. C. Deb., Standing Committee D, col. 616 (9 June 1981).
15 E.g. Pleasure Craft (Arrival and Report) Regulations 1979, S.I. 1979 No. 564, reg. 2; Tor Bay Harbour Act 1970, s. 4(1); Chichester Harbour

Conservancy Act 1971, s. 86(9); Crouch Harbour Act 1974, s. 66(9); Eastbourne Harbour Act 1980, s. 18(3).
16 H. L. Deb., vol. 424, col. 529 (15 October 1981).
17 Control of Pollution Act 1974, s. 31.
18 Wildlife and Countryside Act 1981, Sch. 12.
19 Lundy (Bristol Channel) Marine Nature Reserve Order 1986.
20 Territorial Sea Act 1987, s. 3(2)(b), Sch. 1, para. 6.
21 Wildlife and Countryside Act 1981, s. 36(1).
22 Section 5.
23 Section 36(6).
24 Clause 3.
25 H. C. Deb., Standing Committee D, cols 12–16 (6 March 1985).
26 Section 3.
27 Schedule 12, para. 7.
28 Department of Transport. See *London Gazette*, 27 November 1986.
29 Wildlife and Countryside Act 1981, s. 36(1).
30 Article 1 (2), Treaty Series No. 48 (1982), Cmnd 8714.
31 H. L. Deb., vol. 415, col. 994 (16 December 1980); H.L. Deb., vol. 416, col. 1178 (3 February 1981); H.C. Deb., Standing Committee D, cols 578, 600 (9 June 1981); H.C. Deb., vol. 8, col. 930 (13 July 1981).
32 *International Legal Materials* 19 (1980) 15.
33 Treaty Series No. 56 (1982), Cmnd 8738.
34 Section 15.
35 National Parks and Access to the Countryside Act 1949, s, 20.
36 Sea Fisheries Regulation Act 1966, s.7.
37 Wildlife and Countryside Act 1981, ss. 28–9.
38 ibid., s. 28(1)(a).
39 See note 25 above.

REFERENCES

Department of the Environment (1981) *The Establishment of Marine Nature Reserves*, consultation paper, London : HMSO.
House of Commons Environment Committee (1984–5) *First Report from the Environment Committee: Operation and Effectiveness of Part II of the Wildlife and Countryside Act*, H.C. Paper 6–1.
Nature Conservancy (1968) *Conservation Policy in the Shallow Seas*, London: HMSO.
Nature Conservancy Council (1986) *Skomer Proposed Marine Nature Reserve*, consultation paper, London: HMSO.
Nature Conservancy Council (1988) *Menai Strait Proposed Marine Nature Reserve*, consultation paper, London : HMSO.
Nature Conservancy Council/Natural Environment Research Council (1979) *Nature Conservation in the Marine Environment*, London: HMSO.
National Environment Research Council (1973) *Marine Wildlife Conservation*, London: HMSO.

Part III

The management of individual uses: mineral resource development

7 The management of UK marine aggregate dredging

F.G. Parrish

Although its roots can be traced back several centuries, the UK marine aggregate industry as we know it today is almost exactly twenty-five years old. In its fairly short life the industry has already seen rapid expansion and then in the early 1970s a decline in the fortunes of the building industry generally. Over the last few years we have seen a steady increase in the landings of marine materials to record levels and we are now seeing significant new investment in ships, prospecting and wharf facilities for the future. At the same time the focus of environmental concern on the marine environment and activities in it mean that those managing and controlling the industry have to be responsive to such concern if we are to see the conditions in which the industry can continue to thrive. Few other countries have a similar industry – the UK is fortunate in possessing offshore resources accessible to markets in a way which no other country, in Europe certainly, does. The present industry has been called a 'gatherer' industry, but it contains within it the prospect of practical technological and systems development which can put the UK among world leaders in a field in which there is considerable international interest.

THE INDUSTRY

Most of the tonnage landed from the UK territorial sea and continental shelf is dredged by four main companies – United Marine Aggregates, Ready Mixed Concrete, Civil & Marine and ARC Marine. Between them they own about 80 per cent of the capacity of the present aggregate dredging fleet and land about 85 per cent of total annual extraction. The total fleet at present is about fifty ships, which landed in 1989 just under 21 million tonnes for the UK aggregate market. There are also a number of smaller, locally based, independent companies. In addition to this, contract dredgers, mostly Dutch, also extract marine materials for both landfill and beach replenishment – very large examples being the highly successful Seaford beach replenishment project in 1987 and the Channel Tunnel entrance works at Cheriton in 1989.

Most of the ships dredging at present date from the early to mid-1970s and range in capacity from about 600 tonnes to 8000 tonnes, with the majority at about 4000 tonnes. The great majority are trailer suction dredgers, although there are still a few smaller vessels, particularly on the west coast, which carry out anchor dredging.

Overall, marine aggregates provide about 17 per cent of the UK's total requirement of sand and gravel. The important markets are in the south and south-east – particularly the London area. In 1989 over 8 million tonnes, about a third of the total dredged, went into the Greater London and south-eastern area, meeting about 30 per cent of the total requirement for the area. On the south coast, Hampshire and Sussex depend on marine aggregates for about 50 per cent of their total requirement. The importance of the industry nationally, regionally and in terms of shipbuilding and employment is therefore substantial.

MANAGEMENT

The only statutory consent required for aggregate dredging outside port areas is that of the Department of Transport under the Coast Protection Act 1949. Beyond this, management and licensing arrangements rest on the common law powers of the Crown Estate as owner of the resources. It is as well here to explain something about the Crown Estate itself: it is not a government department in the commonly understood sense and its role is not administrative. The Crown Estate Act 1961 makes the Crown Estate Commissioners a body corporate, charged on behalf of the Crown with managing and turning to account land, etc., held in right of the Crown. Under the Act the Commissioners' general duty is to maintain the Crown Estate as an estate in land, to maintain and enhance its value and the return obtained from it but 'with due regard to the requirements of good management'. (Reference Crown Estate Act 1961).

The 'Crown Estate' is an accumulation of land and property belonging to Her Majesty 'in right of the Crown'. Since 1760 the annual revenues have been passed to Parliament – the Consolidated Fund – and in return Parliament undertakes those costs of government which previously fell on the Crown's income, and grants the Civil List payment. In the year ending March 1989 the crown Estate's contribution to the Exchequer was £41 million, of which about £2.5 million came from aggregate dredging activities.

The foreshore and bed of the sea are a very ancient Crown possession. The Crown Estate still owns roughly 55 per cent of the foreshore – the land between mean high and mean low water, most of the bed of the territorial sea (now twelve miles), with very few exceptions, and beyond the territorial sea limit the right to explore and exploit the natural resources of the continental shelf are also vested in the Crown under the management of the Commissioners. In all cases, of course, oil, gas and coal are excluded, as they are from estates on land.

It is interesting to note that the major expansion of aggregate dredging in the early 1960s coincided almost exactly with the introduction of the Continental Shelf Act 1964, which put beyond doubt the Crown's ability to grant legally enforceable concessions for sand and gravel extraction.

THE LICENSING SYSTEM

The commissioners require that before considering an application to extract sand and gravel from the sea bed the prospective applicant must satisfy them that the company has the resources and expertise to meet the licensing conditions and that the applicant should take a prospecting licence.

Prospecting licences can cover fairly small areas of sea bed: they generally run for a period of between three months to two years and permit the holder to carry out prospecting by seismic survey, sidescan sonar, grab samples and boreholes. In addition the licence generally permits a limited quantity to be taken by dredger sampling. The normal limit is 500 tonnes over the period. In the past dredger sampling was a very common method of prospecting and it remains important to dredging companies in order to assess the commercial value and viability of the cargoes which can be dredged.

Once a resource has been identified and prospecting carried out in accordance with the terms of a prospecting licence, the applicant may apply for a production licence. As mentioned above, there is no statutory basis beyond the powers of the Department of Transport, which relate solely to navigational interference, under which an application can be assessed. What has therefore evolved is an informal consultation procedure under which the commissioners have agreed that they will grant licences only if, following consultation with the relevant government departments, there is no substantive objection to the proposal.

The first step in considering an application is to refer the proposal to Hydraulics Research Ltd (HRL) - previously the Hydraulics Research Station of the Department of the Environment – to advise whether there is any likelihood of the dredging causing damage to the adjacent coastline. The information given to HRL is a chart, showing by co-ordinates the proposed area of extraction, the prospecting report and a proposed annual maximum extraction rate, normally based on 5 per cent of the likely reserves.

In forming an opinion on the licence application HRL aim to answer the following questions:

1 Is the area of dredging far enough offshore so that material is not drawn down the beach into the deepened area?
2 Is the area to be dredged sufficiently far offshore and in deep enough water so that changes in the wave refraction pattern do not take place? Such changes may alter the longshore transport of beach material and hence affect shoreline stability.
3 Does the area of dredging include offshore bars which are sufficiently high to give protection to the coastline from wave attack? A significant reduction in the crest height might increase wave action at the shoreline and lead to erosion.
4 Is the dredging to be carried out in deep enough water so that it will not affect possible onshore movement of shingle?

After an initial assessment of the proposal, HRL inform the Crown Estate of the extent of research they consider necessary. This may include desk study, computer simulation, site inspection or in some cases a full-scale field research programme. The Crown Estate requires the cost of this study to be met by the dredging company, which must decide on the basis of the information provided by HRL whether the cost of the research required is economically justified. If the company feels unable to meet the cost of the research considered necessary the application proceeds no further.

In consulting HRL the Crown Estate seeks an assurance that the dredging proposed will have no significant affect on the coastline – the 'zero effect principle'. Given that we are dealing with a low-cost commodity and a high level of uncertainty whether any production application will be granted, it is well understood by the companies, the Crown Estate and HRL that what is required is the highest measure of surety at the lowest economic price. The result of these requirements is that HRL's judgements err well on the side of caution. The computer model itself tends to overestimate effects, and the requirement of zero effect means that the consultant's views will be based on the worst case. If, on this basis, HRL's view is that there would be deleterious effects on the coastline the company is informed and the application goes no further. Before reaching that stage, however, it is common for HRL to discuss their potential results with the Crown Estate and its advisers and with the company concerned to see whether by placing restrictions on the proposed dredging (e.g. a depth limitation on the lowering of the sea bed allowed), or by sponsoring further and more expensive research, the difficulties might be overcome.

Presuming that HRL offer a favourable view and that the Crown Estate's own checks on such issues as existing pipelines, cables, etc., have shown no obstacles, the application, supported by HRL's report, the prospecting report and the Crown Estate's own views, are put to the Department of the Environment (DoE) Minerals Division for a 'government view'.

Minerals Division obtains this view by consulting all departments whose interests might be affected by the proposal. It is, in effect, a 'cascade' consultation procedure under which Minerals Division will go initially to:

1 Other divisions within the Department of the Environment (Construction Industries Division, the sponsor for the industry; Rural Affairs Division on environmental aspects).
2 Ministry of Agriculture, Fisheries and Food (fisheries and coast protection).
3 Department of Transport (navigation), which will indicate at this stage whether it will be able to issue a consent under the Coast Protection Act.
4 Ministry of Defence (hydrographic and naval interests).
5 Department of Energy (oil, etc., interests).

Each of these departments will in turn consult other bodies such as local coast protection authorities, fishery committees, the Nature Conservancy Council, regional water authorities water authorities, ports, etc. Some of these bodies will consult further.

The government view effectively rests upon whether or not any government department sustains an objection to the proposal. As soon as a substantive objection has been raised Minerals Division will inform the Crown Estate, which, after discussion with the applicant, sees whether it is possible by negotiation, by further research or by changes in the area or licence conditions to resolve the objection. Thus, although the initial consultation period is limited to a matter of a few months, the final resolution of an application can take several years. If it is not possible to resolve any objection, then a favourable view is not given and a licence will not be issued.

It may be worth while to give some examples of the kinds of objection which have been raised and the ways in which compromise solutions have been sought and found. The most common grounds for objections, and the ones where there is most scope for compromise, are coast protection, fisheries and navigation.

COAST PROTECTION

The HRL report which accompanies every application seeking a government view forms the basis of all discussion on coast protection issues. Once they have given a favourable view in their report, HRL will act as independent consultants in considering any further objections. HRL's research has shown that, as a general rule of thumb, dredging below the 18 m depth contour does not produce discernible effects on nearby coastline. Even then, however, objections can sometimes lead to modification of an application, for example by restriction on the dredging depth permitted or by excluding crest areas which might afford some protection for inshore spending banks. In other cases it is possible by further research to demonstrate that dredging can safely be permitted inshore of the 18m contour, for example subject to annual bathymetric surveys and further monitoring by HRL of the progress and effects of dredging. It should be noted that although the hydrodynamic studies are carried our economically they are nevertheless based upon HRL's considerable expertise and experience in this field and carry the full weight of their national and international reputation.

NAVIGATION

Generally, navigational aspects are fairly straightforward. Initial objections can sometimes be raised as a result of misunderstanding about the manoeuvrability of dredgers, and these can be comparatively easily dealt with. In some cases a specific requirement can be inserted in the licence stipulating that only trailer dredging may be carried out and that no dredging will be carried out at anchor. In other situations the reverse might apply. A common requirement is that dredging shall take place in line with the general flow of shipping traffic in an area. Apart from direct interference with navigation, there might be fears about the dredging affecting navigational channels. Where necessary, further studies by HRL can validate or remove such objections.

FISHERIES AND THE ENVIRONMENT

The interrelationship between fishery and dredging is probably the one which causes greatest difficulty. It demands close consultation between the Crown Estate and MAFF on all dredging proposals. In addition to the government view procedure, therefore, a code of practice was agreed with MAFF, the terms of which were published in December 1981 (MAFF 1981). The aim stated in this document is:

> The purpose of this Code of Practice is to provide a basis for close liaison at working level between the fishing and dredging industries in order to promote mutual co-operation and to reduce to a minimum potential interference with each other's activities and damage to each other's resources.

Because under the present procedure prospecting licences do not give a company exclusive rights to an area, prospecting applications are considered commercially confidential. Before a prospecting licence is issued, however, the Crown Estate informs MAFF headquarters in confidence of its intention to do so and MAFF consults its fishery laboratories and the local district inspector of fisheries. They then inform the Crown Estate and the company whether there are likely to be conflicts in the area in question. Once a prospecting licence has been issued the Crown Estate sends details to MAFF headquarters and the district inspector and at that stage the local sea fisheries committee is also informed. The dredging company is also put into contact with the district inspector and the sea fisheries committee so that, right from the start, working contacts can be established and problems discussed. In particular any proposals for dredged samples require clearance by the Crown Estate, and in each case MAFF is consulted on the timing and the area concerned. In this way the effects of sampling on, for example, spawning grounds or shellfish beds are minimized.

At the production application stage, under the government view procedure, MAFF, as explained above, consults its laboratories, the district inspector and the sea fisheries committee which co-ordinates local fishery interests. Under the code of practice, where MAFF intends to object to a licence, it notifies the Crown Estate and the company, explaining its concern. Invariably these prove both contentious and time-consuming – if they can be resolved at all.

Many difficulties arise from the paucity of information on a local scale on the volume, type and quality of fishing in particular areas. By the time a dredging application gets to a production application stage we are talking typically of an area of less than 50 km^2, in some cases less than 20 km^2; ICES squares on which fishery information is generally available are typically 1100 km^2 in extent. The interaction between dredging and the environment is complex and does not fall within the purview of this chapter; in general terms, however, it is well accepted that while there will be some damage to commercial fish stocks from the dredging operations itself these are small and, except in unusual circumstances, insignificant.

Each application is considered on its merits, taking into account the extent and type of benthos, the possibility of change in the sea bed environment, the existence of spawning grounds for herring and sand eel, shellfish beds and fixed-gear fisheries.

In a chapter of this scope it is impossible to cover the range of interactions or the scientific studies which have been carried out. Unless, however, MAFF can be satisfied that the proposed dredging will not have a significant effect on commercially important fish stocks or their food chain a favourable government view will not be granted.

The question of the environmental effects of dredging is often confused and overlaid with passionate arguments about physical disturbance by dredgers to established fisheries. Clearly, in considering whether to raise an objection, MAFF has regard to the size and the commercial value of the fishery concerned. Without doubt, some fishery activity will be affected by the dredging activity. In general, however, with good co-operation and prior information in accordance with the code of practice, the disturbance can be minimized, and the cases of direct conflict within established dredging areas are very few indeed. Trawlers can work quite successfully in the close vicinity of dredgers. Fixed-gear fishermen clearly cannot work in an area which is subject to regular dredging but again the code of practice is designed to ensure that information on dredging activity in a particular area is passed regularly to the local district inspector and sea fisheries committee so that areas which are not actively being dredged can be used by fishermen.

It is important to remember too that, while dredging grounds may look extensive on Admiralty charts, the actual area being dredged at any one time is comparatively small – for example, if one assumes an average dredging depth of 1 m the area of sea bed actually touched by dredging pipes in a year amounts to about 13 km^2, a tiny proportion of the total UK territorial sea and continental shelf. Within this, of course, there is considerable scope for intense local conflict but it is our firm view that, given good communication between dredging and fishing industries at all levels, such conflict can be resolved, or at least ameliorated.

PRODUCTION LICENCES

Only when all objections have been resolved and a favourable government view has been affirmed is a production licence issued. The licence from the Crown Estate stipulates, by reference to an extract from the Admiralty chart and co-ordinates, the area in which dredging is permitted: it lays down an annual permitted maximum and it also lays down any conditions required by the government view itself. In a recent case, for example, the government view conditions required that the area could be dredged only by suction trailer dredgers, that the area should be buoyed, that only those areas within the licensed area where it had been proved that a sand and gravel thickness of over 1 m in depth existed could be dredged, that a capping layer of sandy gravelly material of not less than 0.45 m was to remain after dredging, and that in no circumstances was the overall depth of the dredging to exceed 2 m. In other cases there have been requirements of seasonal adjustment, or night dredging only.

The licence also stipulates the annual royalty rate. Royalties are payable per tonne of material extracted from the sea bed. Returns are made twice a year by the dredging companies in the form of a statutory declaration. In order to discourage large areas being licensed and sterilized by companies there is also a provision for a 'dead rent' of 20 per cent of the royalties payable on the permitted annual take.

The licence also requires that the company will give the Crown Estate full access to its records, including log books, etc., and that its accounts may be audited by the Crown Estate.

MONITORING OF LICENCE COMPLIANCE

The Crown Estate, like any landlord, is concerned to ensure that the terms of the licence are strictly complied with. We can and do audit companies' records and we investigate all complaints of unauthorized dredging. In investigating such complaints we can require access to all company records, including the vessel's log, and we get very good co-operation from coastguards and port authorities. We can and do where necessary interview the masters of the vessels and the companies' operations staff. In some cases we carry out sidescan sonar surveys of the area, and these can readily identify whether dredging has been carried on outside the permitted zones. MAFF's fishery patrol spotter aircraft are tasked to report all dredgers sighted in the course of their patrols, and they provide a formal statement if any dredger is seen dredging outside its limits. The Crown Estate Commissioners have consistently taken a strong line in the event of unauthorized dredging being proved. The sanction available to the commissioners is to curtail the licence in question, either permanently or temporarily, and this has been done in the past. In some cases the companies themselves have also sacked or suspended masters of vessels. We believe, however, that unauthorized dredging is an infrequent occurrence nowadays; large companies with large investments do not risk the licences on which their livelihood depends.

THE SITUATION NOW

After about twenty-five years what we have now is approximately 90 licences around the British coastline issued to a total of twelve companies, mainly concentrated in the southern North Sea, along the south coast from about Worthing to Christchurch, in the Bristol Channel and on the north-west coast (see Figure 7.1). As mentioned above, the largest quantities are taken from the east coast. Few licences, apart from those required for short-term use for coastal contracts, have been issued over the last ten to fifteen years and it is now most important that new resources are released if the industry is to have an assured future. The Crown Estate, the Department of the Environment and the companies are working together in reviewing all aspects of the present government view procedure and the licensing system. As I write we are in the midst of these reviews and it is possible to pick out a few important strands and indicators of the way in which management will develop in future.

The EC Directive on Environmental Assessment will cover the offshore area as well as land. Under the directive any application for mineral activity which is likely to have a significant effect on the environment will have to be accompanied by an environmental impact assessment. The DoE's consultation paper on the implementation of the directive proposes in essence that this requirement will be incorporated into the government view procedure. The DoE's consultation paper on the revisions to the government view procedure itself is awaited. Criticisms of the present system which have been raised by the dredging industry are:

1 There is no single government authority responsible for developments in territorial waters and the UK continental shelf. Many departments and organizations have interests in the sea bed and separately keep information relevant to their interests. This may lead to unjustified refusals or unnecessary restrictions being placed on dredging.
2 There is no overriding authority or mechanism within government with responsibility for deciding a particular application on its merits. If one department maintains an objection it is sufficient for an application to be refused, whatever the merits of the case.

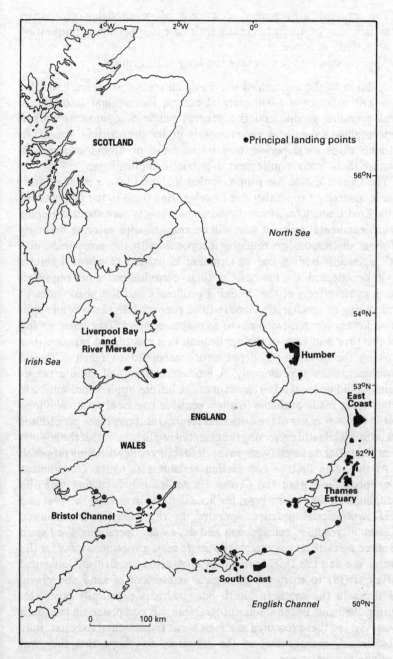

Figure 7.1 Principal dredging areas

3 The present arrangement does not allow dredging companies sufficient opportunity to explain their case to objectors or to question objections in public.

4 Applications for licences take too long to determine.

In addition to the concern of the dredging companies there have also been criticisms from environmental bodies, recreational associations, local pressure groups and the general public who have no formal opportunity to comment on proposals under the existing system. In extreme cases local people may be unaware of dredging proposals because there is no requirement to publicize applications.

The Crown Estate has proposed that it should take a more positive role in assessing proposals prior to submitting them to the Department of the Environment for a formal government view. We are about to employ an environmental scientist who will be charged with carrying out early informal discussions on dredging proposals with the companies, and with agencies, bodies and government departments whose interests might be affected. On the basis of these consultations a full report on the possible effects of the dredging proposals and the possibilities of overcoming or ameliorating them will be prepared. In this way an application for a government view can be accompanied by a full statement setting out the pros and cons of the application in a considered manner, thus allowing the government department concerned to reach a formal government view expeditiously. A fundamental part of these arrangements would be local advertisement of the licence application, sufficient detail being made available to allow sensible comment on it. We hope that in this way many of the criticisms levelled at the present procedures can be met while still preserving the essential requirements that the government should take a formal view on the desirability of any dredging proposal.

At the same time as considering amendments to the Government View procedure itself the Crown Estate is also discussing with the dredging licensees a new basis for licensing, incorporating the best and most professional practice operated in the industry on resource assessment, resource management and day-to-day operations. We regard resource assessment as possibly the single most important aspect for the future. We and the DoE have jointly contracted the British Geological Survey (BGS) to carry out geological assessment of sand and gravel resources in the southern North Sea, which is practically complete. During 1988 and 1989 a similar programme was completed on the south coast. We see these resource assessments as fundamental to better, fully professional, management of the resource, and hope that they will continue in other areas.

We are also exploring with the companies a more positive system of prospecting licences which will not only give the companies added assurance at a time when they are committing considerable expenditure to commercial exploration but will also permit much wider and more open consultation at the earliest stages of licensing.

We are also considering with the companies whether, as technology becomes available at an economic cost, electronic monitoring systems might be introduced for dredging vessels in the future. This on its own is a topic for a discussion paper. In general, however, it is our view that, subject only to navigational systems which at reasonable cost can produce and record accurate position fixes, it would be in the interests of the dredging industry to introduce such a system. Not only would it produce detailed management information for both the companies and the Crown Estate but with such a system investigations into allegations of unauthorized dredging could be carried out and the position established beyond reasonable doubt. At present the Decca systems generally used by the industry do not give this level of accuracy. We hope, however, that the industry will co-operate with us in introducing some such system as and when the technology becomes generally available.

In addition to these particular proposals we are discussing with the companies a wide range of issues, including requirements for prospecting standards and an approach to licence applications more akin to that required for mineral planning applications on land. Clearly the circumstances are not directly comparable; the costs and the difficulties of prospecting and operating at sea are much greater, and due allowance must be made for this. We believe, however, that the industry has now reached the stage when a more positive approach to the management of the resource by both the Crown Estate and the companies will pay dividends in terms of future security and progress for the dredging industry as a whole.

NOTE

Many of the figures quoted in this chapter are now out of date. In 1989 about 21 million tonnes were dredged for the UK aggregate market. The references give the up-to-date position on the Crown Estate's policy on marine aggregate dredging and the Department of the Environment's revised government view procedure.

REFERENCES

Ministry of Agriculture, Fisheries and Food (1981) *Code of Practice for the Extraction of Marine Aggregates*, London.

The following publications, which appeared after the conference, summarize management changes.

Crown Estate (1988) *Licensing and Management of Marine Aggregate Dredging: Policy Paper by the Crown Estate*, London: Crown Estate.

Department of the Environment and Welsh Office (1989) *Offshore Dredging for Sand, Gravel and other Minerals*, London: Department of the Environment, and Cardiff: Welsh Office.

8 Research and technology requirements for the evaluation of the resource potential of ridge crest deep-sea polymetallic sulphide deposits

John Yates

This chapter is based on recent work by the author aimed at formulating a development scenario for deep-sea polymetallic sulphide deposits, which will both maximize the benefits to basic science and provide the underpinning research necessary prior to serious consideration of commercial exploitation. It begins with a brief description of the nature, genesis, average grade and abundance of MPS deposits based on the data so far collected by the MPS programmes in the United States, Canada, France, the Federal Republic of Germany and Japan. This is followed by the results of a large international postal survey of key academics, designed to establish a consensus of research needs in the area, and the results of a similar exercise to establish the views and requirements of industry. The results are presented following the normal chronological sequence for the exploitation of any resource; location/survey, sampling/proving, extraction and processing. The chapter concludes with a consideration of the economic, legal and environmental factors which will influence the viability of MPS deposits as a resource.

INTRODUCTION

Oceanographic research during this century has revealed an increasing number of marine mineral deposits. One of the most recent of these discoveries has been marine polymetallic sulphides (MPS). These deposits are high-grade and localized, forming in areas of active undersea volcanism, normally at a depth of 2000–3000 m. Their discovery in the late 1970s has been heralded as one of the most important scientific findings in marine science this century. The areas which host MPS deposits have provided geologists with an ideal opportunity to study primary ore-forming hydrothermal systems in real time. In addition, these systems support a unique and varied fauna, which depend upon this volcanic activity for their existence. Finally, the quantity of minerals added to sea water by this volcanism has radically altered previous ideas about the factors which contribute to the chemical composition of the oceans.

Polymetallic sulphides are so called because of the wide range of minerals they contain (up to twenty). Many samples recovered so far have average grades which if present in sufficient quantities on land would constitute a valuable resource. This has fuelled speculation that MPS may one day provide an alternative source of metals to traditional land-based supplies. However, at present not enough is known of the true abundance of the deposits, nor of what represents an 'average' deposit. Nevertheless results to date have been encouraging and, as more high-grade deposits are discovered, commercial interest in MPS is likely to grow. It is therefore timely to consider what technological alternatives exist or are likely to be developed which will enable commercial extraction to take place.

Deep-sea mining is no longer confined to the drawing board. Several sophisticated systems have been developed with a view to extracting manganese nodules at depths of 4000–6000 m, and a project in the Red Sea has proved the viability of pumping metalliferous muds from over 2000 m depth. Many elements of these systems may well be applicable to MPS extraction, but one major problem remains to be overcome. Unlike the Red Sea muds and manganese nodules, MPS are hard consolidated deposits, similar in form to land-based sulphide ore bodies. This means that any system for mining them will have to include a mechanism for disaggregating the deposit prior to transport to the surface, and at present the technology for achieving this in the deep ocean does not exist. Alternatives have been suggested, such as solution mining or vent capping, but this chapter will show that physical crushing of the ore is still viewed as the most likely solution by most engineers who have considered the problem. Current proposals are still at a very rudi- mentary stage of development and it is likely that the ideas will be substantially refined as the knowledge base on MPS continues to grow.

Technological requirements are not the only criteria which will dictate whether or not MPS ever becomes a viable economic resource. Consideration must also be given to non-technological factors: world demand for the metals they contain, the status of deep-sea deposits in international law and environmental considerations.

Genesis and abundance

It is important to stress that the current data base on MPS is very small. The vast majority of deposits found to date have been located on active spreading centres, which extend for some 50000 km across the floor of the world's oceans. However, only around 150 km of this system has yet been investigated in sufficient detail to reveal any MPS deposits which exist there. In addition, recent evidence suggests that back-arc rift zones and associated geological regimes may also host sulphide deposits. If so this would greatly increase the global potential for MPS formation.

MPS deposits are the product of hydrothermal systems. Figure 8.1 shows a schematic diagram of such a system at a spreading centre. Sea water enters the deep fissures present at ridges due to tectonic movement and thermal contraction of the rock, and rapidly becomes heated by the high geothermal gradient. This water may descend to depths of up to 5 km, and be heated to 300–600°C (Edmond 1983: 491-2). As it percolates through the hot rocks it leaches out certain components, including transition metals, silicon, calcium, potassium, hydrogen and sulphur. The superheated metal-laden brines are much less dense than unheated sea water and rise to the surface wherever possible. On reaching the sea floor, still at temperatures of up to 350°C, they emerge into near freezing abyssal water (average temperature 2°C). This rapid cooling produces a drastic reduction in solubility, and sulphide and oxide minerals precipitate out as a cloud of particles, producing an effect known as 'smoking'. These vent emissions tend to be darker in colour as the temperature rises, leading to a subdivision between 'white smokers' (relatively low-temperature), and 'black smokers' (high-temperature).

The minerals so produced build up into 'chimneys' around the emitting vents, spire-like structures which may grow to 20 m or 30 m high. These chimneys are inherently unstable owing to the friable, inhomogenous nature of the sulphides and are further weakened by the effects of oxidation and the active seismicity endemic in these areas. Consequently they frequently topple over, leading to an accumulation of deposits composed of chimney fragments, known as the basal mound (Haymon 1983). In addition it is likely that mineralization is widespread below the sea floor in the form of a stockwork (a network of mineralized veins), though, as will be shown later, information on the vertical dimension of the deposits is extremely scarce.

There are five basic elements necessary for the formation of a hydrothermal ore deposit: a source of metals, permeable pathways through the source rock, heat, water, and a method of triggering precipitation to form an enriched deposit. Spreading centres provide an

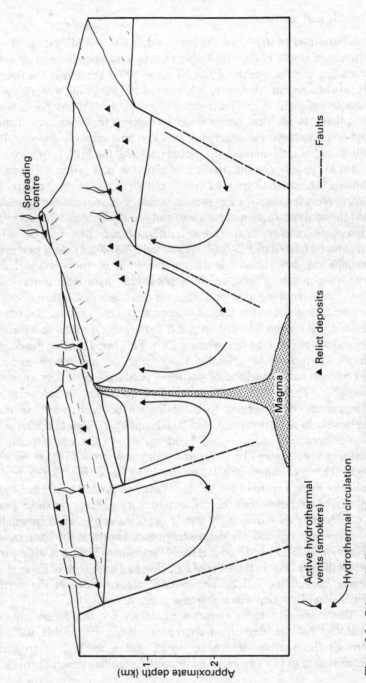

Figure 8.1 Schematic diagram of a submarine hydrothermal system

abundance of all five factors. The basalt of the sea floor provides a rich source rock for transition metals and, as mentioned previously, is invariably fissured around oceanic ridges, providing permeable pathways. Heat is provided by the high geothermal gradient, and water by the overlying 2000–3000 m of sea water. Finally, precipitation is initiated by the temperature difference between the existing fluids and the ambient sea floor temperature. So the necessary ingredients of a hydrothermal system occur along the whole length of the oceanic ridge system. In addition recent finds have indicated that MPS deposits may well also form in back-arc basins, again in association with undersea volcanism and rifting (Uyeda 1987: 73). Not only do these sites contain all the vital prerequisites necessary for ore-formation, but it has been postulated that they have the potential to produce even richer deposits (Sillitoe 1982).

The first discovery of a sizeable submarine deposit of metalliferous sulphides was made in the Red Sea. Metal-rich muds were found along the axis of the spreading centre bisecting the Red Sea, where they accumulate in 'deeps' (depressions in the sea bed where dense brines of hydrothermal origin can accumulate at temperatures which may exceed 55°C). The Atlantis II deep has been the subject of detailed study since 1966, and has proven reserves estimated to be capable of producing over $2 billion in revenue over a sixteen-year operation (Mustafa *et al.* 1984). In 1974 the governments of Saudi Arabia and Sudan concluded a bilateral agreement forming the Red Sea Commission (RSC), with a view to exploiting the Atlantis II deposit on a commercial basis. The RSC engaged the French Bureau de Recherches Géologiques et Minières (BRGM) as geological consultants and the German company Preussag as technical contractors. Preussag subsequently developed a system which is capable of pumping the muds to the surface from a depth of 2200 m, and it is possible that commercial extraction will commence within the next decade.

The Red Sea deposits appear anomalous in that they take the form of unconsolidated muds, unlike the massive consolidated MPS deposits forming chimneys and mounds on spreading ridges in the open ocean. Sizeable deposits of this type have now been located at various sites world-wide (see Figure 8.2), the principal ones being on the Galapagos rift (00°45′N, 85°50′W), the East Pacific rise (21°N; 20°S; 13°N), the Explorer ridge (49°454′N, 130°15′W), the Juan de Fuca ridge (45°59′N, 130°04′W; 47°57′N, 129°06′W) and the Gorda ridge (41°00′N, 120°30′W; 40°45′N, 127°30′W). Many smaller finds have been made, including recent discoveries in back-arc settings, and slow spreading centres (Rona *et al.* 1986). The fact that the major discoveries to date

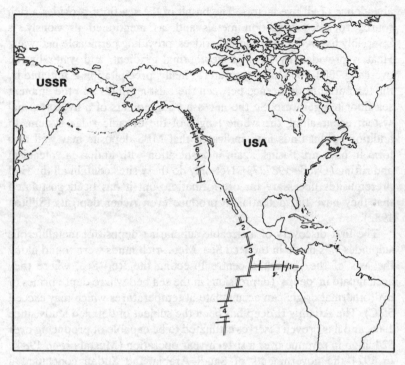

Figure 8.2 Major MPS deposits. *1* Galapagos rift, *2* EPR 21° N, *3* EPR 13° N, *4* EPR 20° S, *5* Explorer ridge, *6* Juan de Fuca ridge, *7* Gorda ridge

have tended to be in the eastern Pacific is almost certainly because this is the area which has been subjected to the most detailed survey work and not because they possess characteristics which make them preferential sites for mineralization.

The grades of ore found at these sites have varied enormously, not only from locality to locality but also between samples taken from the same site. This highlights the first major requirement for an accurate resource assessment to be made, more data on metal content from known sites coupled with increased exploration for new ore bodies, so that a picture of what constitutes a 'typical' MPS deposit can be built up. Table 8.1 depicts results from sampling at a variety of localities to give a flavour of the range of grades to be expected. To date the most promising constituents appear to be copper, zinc, cobalt and silver, but as more samples are collected it is possible that this assessment may need to be revised in the light of incoming data.

Table 8.1 Typical contents of selected metals from sea floor polymetallic sulphide samples and related ancient massive sulphide ores

Location	21°N Active vents	CYAMEX area	Juan de Fuca ridge			Galapagos rift 86°W	Guaymas basin	Cyprus Ores	Ancient Ores	Canadian Precambrian
			Northern	Southern	Axial seamount					
Zinc (%)	32.3	40.8	6.3	54.0	19.2	0.1	3.8	0.2	0.5	5.0
Copper (%)	0.8	0.6	0.5	0.2	0.13	5.0	0.4	2.5	4.0	2.0
Lead (%)	0.3	0.05	0.1	0.3	0.4	0.0	2.1			
Silver (ppm)	156	380	30	260	288	10	477	39		50
Gold (ppm)	0.17	0.08	0.13	0.05	0.3		1			

Source: after Scott (1984).

RESEARCH PROGRAMMES AND REQUIREMENTS

The study of sites of active and relict hydrothermal activity has been both multi-disciplinary and international in nature. The majority of dedicated expeditions have been launched by government-funded agencies from various countries. In the United States the main participants have been the National Oceanic and Atmospheric Administration (NOAA), the United States Geological Survey (USGS), the Minerals Management Service (MMS) and the United States Bureau of Mines (USBM). In the Federal Republic of Germany the major government funding has been channelled through the Bundesanstalt für Geowissenschaft und Rohstoffe (BGR), Bundesministerium für Forschung und Technologie (BMFT) and Deutsche Forschungsgemeinschaft (DFG). Canada has an active programme in MPS exploration primarily funded by the Ocean Mining Division of the Canada Oil and Gas Lands Administration (COGLA). The main thrust of the French programme is carried out by the national oceanographic agency, Institut Français de Recherche pour l'Exploitation de la Mer (IFREMER), with contributions from BRGM and the marine mining consortium GEMONOD (Groupement d'Interêt Public pour la Mise au Point des Moyens Necessairs a l'Exploitation des Nodules Polymetalliques). Finally, Japan has a growing programme in MPS exploration, orchestrated by the Japan Marine Science and Technology Centre (JAMSTEC), and the Ocean Research Institute (ORI) of the University of Tokyo.

In addition to the above agencies many universities and oceanographic institutes have contributed to knowledge of MPS deposits. These include the Woods Hole Oceanographic Institute and Scripps Institute of Oceanography in the United States, much of their funding being provided by the National Science Foundation (NSF), Bedford Institute of Oceanography (Canada) and the Institute of Oceanographic Sciences (UK). The USSR has also recently initiated a programme of research into MPS deposits and vent-related phenomena, and appears eager to increase co-operation with western scientists (Siapno 1987). Other countries which have shown some interest in MPS deposits include Norway, Australia, New Zealand and Iceland.

Industrial interest in MPS deposits has to date been limited, with the notable exception of Preussag, who have conducted several cruises designed to locate and sample MPS deposits. The study on which this chapter is based found that industry generally is content simply to maintain a 'watching brief' on MPS deposits until such time as sufficient data are available to allow accurate resource assessment. Attempts to establish a consortium in the USA with a view to extracting MPS deposits have met with little interest, and consortia established in the US, the UK, France, Japan and Norway with a view to extracting manganese nodules have at the time of writing devoted no appreciable resources to the assessment of MPS.

MPS deposits possess several advantages over nodules which may make them a more attractive proposition for the private sector. These include:

1 Average grades are far more concentrated than those found in manganese nodules.
2 MPS tend to occur at approximately half the depth of manganese nodules (2500 m compared to 5000 m for nodules).
3 Some MPS deposits are known to occur within EEZs whereas the major nodule deposits are all found in international waters.
4 The processing of MPS ores is likely to be far less problematic than the equivalent for nodules.

Despite these apparent advantages over nodules, industry is extremely wary of committing hard cash to further investigations. This is, in no small part, due to the fact that hundreds of millions of dollars were spent on nodule research, which has still produced no commercial return.

A strong consensus emerged from the survey concerning the factors required to stimulate industry to take a more positive stance concerning MPS, such as applying for exploration leases and developing systems for locating and proving MPS. Three major criteria need to be met before this is likely to occur:

1 More data need to be available concerning the distribution, grades and general geology of MPS.
2 Metal markets must improve, making extraction a more attractive commercial proposition.
3 A predictable and stable legal regime needs to be formalized, relating specifically to the extraction of these deposits.

The main aim of this chapter is to elucidate the factors which will enable the first of the above criteria to be achieved and what technological developments will be needed to extend this basic research work into a development strategy for MPS, if initial results warrant it. Given the rudimentary status of the fundamentals of MPS formation, evolution and morphology, it is essential that on-going programmes of basic research be maintained if not expanded.

A detailed research scenario has recently been devised by a working group on marine sulphides formed as part of the NATO Advanced Studies Institute programme (Edmond *et al.* 1987). The group's recommendations regarding basic research were subdivided into four main areas:

1 *Comparative studies*. Studies of land-based sulphides should wherever possible be related to contemporary marine sulphides and vice versa. Also processes operating on ocean spreading ridges should be compared with those at other relevant geological sites such as back-arc basins and rifts.

2 *Experimental studies*, developing instruments and procedures for monitoring physical, biological and chemical changes at active sites in real time.

3 *Physical modelling*, geotechnical properties, thermal stress models, chemical and hydrodynamic models and models of heat flow loss (both rapid and diffuse).

4 *Chemical modelling*, physical chemistry of the minerals of hydro-thermal systems, studies of solid-solution equilibria, development of chemical/hydrodynamic models which simulate the processes operating beneath the sea bed.

However, in addition to such basic research, work of a more applied nature is necessary to enable a reliable resource estimate of MPS deposits to be arrived at. The following section details the main technological developments required to facilitate this work.

TECHNOLOGICAL REQUIREMENTS

This section analyses the technological requirements necessary to allow exploitation of MPS deposits and follows the normal resource development sequence of location and survey, sampling and proving, extraction and processing. The major techniques in the first two of these phases have been defined by Rona (1983) and are presented in Table 8.2.

Location/survey

Most of these techniques can be regarded as mature, well proved technologies. However, refinements are desirable in many cases to adapt existing technologies to the particular requirements of MPS exploration. Only when this is achieved will location of deposits become routine, but some areas worthy of special attention can be highlighted:

1 More widespread use of existing mapping and survey techniques to highlight areas worthy of further investigation. Particular emphasis should be placed on seamounts, back-arc basins and spreading ridges with a sedimentary input, which to date have received relatively limited attention.

Table 8.2 Exploration strategy of closing range to a mineral deposit

Range to mineral deposit (m)	Method
$10^4 - 10^6$	Regional sediment sampling
	• Concentration gradients of Fe and Mn
$10^4 - 10^6$	Regional water sampling*
	Weak-acid-soluble amorphous suspended particulate matter (ferric hydroxides)
	• δ^3He
	Total dissolvable manganese (TDM)
	• Methane (CH_4)
10^3	Bathymetry
10^3	Magnetics
10^3	Gravity
10^3	Long-range side-looking sonar
10^2	Short-range side-looking sonar
10^1	Bottom images
10^1	Near-bottom water sampling
$10^0 - 10^3$	Electrical methods
10^0	Dredging
10^0	Submersible
10^0	Drilling

* Applies to actively accumulating deposits only.
Source: after Rona (1983)

2 The utilization of dedicated research vessels capable of conducting as many as possible of the techniques outlined in Table 8.2 simultaneously.
3 The development of techniques capable of detecting relict hydro-thermal sites.
4 Methods of processing the results of the surveys as rapidly as possible without the need to employ land-based facilities.
5 The development of geophysical/electrical methods capable of detecting relict sites and/or assessing the extent of deposits vertically.

Sampling and proving

Many of the technologies necessary for proving an MPS deposit are at an embryonic stage of development – in particular, techniques for estimating the extent of a deposit in the vertical dimension. Before commercial production is considered, detailed mapping would be necessary, probably involving a combination of electrical methods, grid-pattern drilling and downhole geophysical methods. Technological developments likely to facilitate the sampling and proving of an MPS deposit include:

1 Further incremental developments in the efficiencies of remotely operated vehicles (ROVs) and deep towed vehicles, particularly in the areas of power storage and telemetry.
2 The development of autonomous ROVs capable of operating efficiently for extensive periods on or near the sea bed.
3 Continuing refinement of deep-sea drilling systems capable of drilling into bare rock.
4 The development of efficient portable hard rock drills capable of operating at depths of 3000 m or more.
5 The refinement and development of electrical or other methods with the potential to produce an estimate of the thickness of a sulphide body without drilling.
6 The development of the technology to produce large-scale optical images of the sea bed from 10 m to 100 m above it.

Extraction

Systems for the extraction of MPS deposits are all conceptual at present. A full description of these various concepts is available elsewhere (Yates 1988), but they can be subdivided into two types: hydraulic/airlift systems or radically new designs. The first type are all based on existing technologies developed for the extraction of manganese nodules or the Red Sea muds. Essentially they comprise the elements outlined in Figure 8.3. The surface facilities would be an adaption of existing technologies evolved for offshore hydrocarbon exploitation and manganese nodule extraction. Transport of the minerals to the surface is achieved via a pipestring, utilizing a series of submerged pumps and/or airlift (injection of compressed air into the upper portion of the pipestring to lower the specific gravity). The most problematic part of such a system would be the need for a novel collector head capable of disaggregating the deposit and of crushing and screening prior to transport to the surface.

An alternative is the development of a radically new technology designed specifically for MPS extraction. Preussag have developed an electro-hydraulic grabdredger equipped with high-resolution deep-sea television cameras, for large-scale sampling of MPS deposits. A larger version may have applications to small-scale mining of very high-grade superficial deposits. The second possibility is the development of a fleet of dedicated autonomous ROVs capable of harvesting superficial

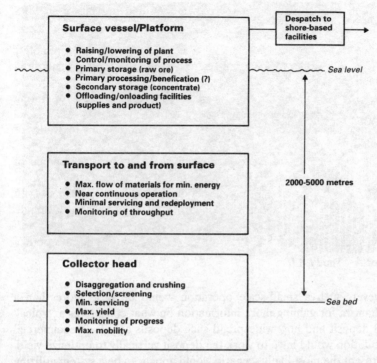

Figure 8.3   Technological requirements of deep-sea mineral extraction

deposits and returning them to the surface without recourse to a pipestring. The French consortium GEMONOD is developing such an ROV, the PLA II (Plate 8.1), with a view to mining nodules, but it is now probable that it will be employed merely as a tethered collector head, ferrying ores to a submerged pipestring. Another possibility which has been considered is 'ventcapping', i.e. the pumping of the emitting fluids direct to the surface for precipitation on board the surface vessel. However, yields would be very low and severe problems would be likely with the blockages of the pipes due to crystallization *en route* to the surface. An alternative might be capping the vents by a detachable cover with an affinity for the desired metals which could be recovered at a later date along with its load of copper, zinc, etc. Finally, solution mining of the deposits has been considered, but present methods rely on highly noxious lixiviants (leaching fluids), and it is unlikely that they would be acceptable on environmental grounds, owing to the problems of containment.

Possibly the most likely first-generation mining vessel would be one employing two or more scaled-up versions of Preussag's grab sampler.

Plate 8.1   The *PLA II*

However, such a small-scale operation would be little more than a mechanism for gaining more information on what constitutes a 'typical' MPS deposit and how widespread such deposits are. True commercial exploitation would have to work the deposit vertically to optimize yield. At present the most likely scenario would appear to be a system utilizing existing hydraulic/airlift technology, combined with a purpose-built collector head, capable of mining the ore by cutting or ripping, followed by crushing prior to transport to the surface in the form of slurry. Such a collector head would be a truly innovative design, and would have to be capable of withstanding hostile environments involving:

1 Near freezing temperatures coupled with the possibility of encountering superheated fluids of several hundred degrees centigrade, which are extremely corrosive.
2 Pressures of 250 atmospheres and more coupled with the possibility of seismic or volcanic events.
3 Extremely rugged topography, containing vertical scarps, fissures, volcanic edifices such as 'lava pillars', etc.

Not only would the system have to cope with all the above problems, but it would also need to be highly manoeuvrable, and extremely reliable and durable. This last factor would be crucial, for the deployment and retrieval of such systems would inevitably be very time-consuming and consequently expensive.

An alternative to the above scenario could be the development of a radical new technological system designed specifically for MPS extraction, such as a shuttle system, vent capping or solution mining. Innovative technological developments would be required in each case, including:

1 High-capacity lightweight power storage systems for autonomous submersibles.
2 Improved methods of telemetry to allow comprehensive monitoring and control of sophisticated submersible vehicles without recourse to an umbilical.
3 The development of an efficient capping mechanism for active vents, and the long-term monitoring of vents to assess the volume and sustainability of emissions.
4 The development of an environmentally safe and efficient lixiviant for use in solution mining.

All the above concepts would require a dedicated, dynamically positioned surface vessel, or semi-submersible platform, but this can now be regarded as an established field of technology and its development should present no real problems. It would be desirable for the surface facility to include some level of primary processing. Again this is a mature technology requiring only incremental changes for use on MPS ores.

Processing

Processing appears to be the least problematic part of any proposed mining operation, as the ores found in MPS deposits are similar to ores mined on land for many years in places such as Canada and Cyprus. Some problems were originally encountered with the processing of the Red Sea muds, owing to the unusually fine-grained nature of the desired minerals, but they have now been overcome. Massive polymetallic sulphides typical of ridge crest deposits are far more similar to land-based ones. Work on the geotechnical properties of samples from the EPR at 21°N has provided valuable information necessary for designing suitable cutting/ripping and crushing equipment, but it is important to bear in mind that all the samples were from superficial deposits and it is possible that sub-surface deposits may differ in their geotechnical properties (Crawford *et al.* 1984).

Primary concentration of the ores would most likely be achieved by some method of froth flotation. The subsequent stages of processing would vary according to the desired metals in a particular sample. Samples from the Juan de Fuca ridge were subjected to chlorine–oxygen

(Cl_2–O_2) leaching with excellent results by the USBM. The process succeeded in recovering 99 per cent of the contained zinc and cadmium, 97 per cent of the silver and 78 per cent of the copper (Sawyer *et al.* 1983). More recently Carnahan (1985) has suggested a standard treatment for MPS ores consisting of blending, crushing and grinding, followed by bulk flotation to remove iron sulphide, selective flotation of copper and zinc sulphides, and final smelting the resulting concentrates to metal. Finally, it may be possible to employ bacteria to release the metals contained in the sulphide ores. The ores are generally of a very porous nature, largely owing to the action of worms and other fauna around the sites of MPS formation. Bio-leaching has been applied in certain instances on a commercial scale but tends to be a slow process. The speed of this method can be increased at higher temperatures, giving rise to the interesting possibility that the thermophilic bacteria present at vent sites may be engineered to win metal from the ores at higher temperatures (Harvey 1986).

It would appear, then, that the technology now exists to process MPS ores efficiently without any radical new developments. However, ore processing is a site-specific activity and so a detailed proposal for benefication and processing can be devised only when details of the geotechnical properties, grades and range of desired metals of a given deposit are available. As always, there is room for innovation, and bio-leaching employing indigenous bacteria engineered to win desired metals from the ore is a possibility worthy of further investigation.

NON-TECHNOLOGICAL FACTORS

The preceding sections detail the research and technological requirements necessary to implement a viable development scenario for MPS deposits. It is to be hoped that basic science related to active and relict vent sites will be continued, if not expanded, considering the wealth of information such localities are providing for marine geologists, biologists and oceanographers. Further development of a more applied nature may be undertaken by government agencies, particularly where benefits are likely to accrue in the wider sphere, for example detailed mapping and survey of new areas. However, such development is unlikely to progress much further than basic mapping and sampling without the involvement of private industry. At present this seems highly unlikely, for two reasons, the first economic, the second legal/political.

World metal markets are in a depressed state relative to the post-war boom period which led to overproduction of many metals, including copper and zinc, resulting in falling prices in real terms over the last ten to fifteen years. Although there has recently been some revival in the prices of many metals, it is too soon to assess whether this represents a long-term trend or is just another short-term boom in a market renowned for its unpredictable nature. Unless demand does rise consistently for the relevant metals (copper, zinc, silver, cobalt, etc.), coupled with a concomitant decrease in profitable land-based supplies, it is doubtful that any private enterprise is likely to invest money in R&D on deep-sea mining systems. The one company which has invested money into this area is Preussag, but it has been assisted by generous government grants. The rationale behind the German programme is as much to develop radical marine technologies for the deep sea *per se* as to secure a new source of metal commodities.

A factor which could strongly influence the economics of deep-sea mining is the strategic value of the contained metals. Metals such as cobalt, which is an essential component of certain steels used in high-performance jet engines, are viewed as strategic, i.e. vital to the interests of national security. It is therefore conceivable that if substantial deposits of such metals were found in the deep sea within a coastal state's national jurisdiction they could be mined by the government or by private industry in receipt of heavy government subsidies.

In addition to the economic constraints on deep-sea mining there are ambiguities in international law which militate against commercial operations. At present two legal regimes govern a coastal state's rights regarding the exploitation of sea bed resources, one pertaining to the area within national jurisdiction, the other covering activities in international waters. Coastal states have enjoyed rights of tenure over minerals found on their continental shelf since the Geneva Convention of 1958. More recently the concept of an exclusive economic zone (EEZ), as embodied in the 1982 United Nations Convention on the Law of the Sea (UNCLOS), has gained general acceptance as part of international customary law. The EEZ concept gives coastal states jurisdiction over minerals out to 200 nautical miles from the baseline from which the territorial sea is measured, but this jurisdiction can extend to up to 350 nautical miles from the baseline or 100 nautical miles from the 2500 m isobath if the sea bed can satisfy geological criteria which qualify it as continental shelf. Within this area of national jurisdiction mining can take place subject only to the laws and regulations of that state, provided it does not interfere with traditional freedom of the seas.

According to UNCLOS the sea bed outside any coastal state's national jurisdiction is termed the 'area' and accounts for some 60 per cent of the ocean floor. UNCLOS designates the resources of the 'area' as the 'common heritage of mankind' and declares that they are not subject to national appropriation. The convention created a new international organization, the International Sea Bed Authority, charged with regulating any mining in the area. Operations would be channelled through the operating arm of the authority known as the 'enterprise', and private industry participating in exploitation would have to transfer relevant technology to the 'enterprise' 'on fair and equable terms and conditions'. Such private contractors would also be subject to strict production controls and annual fees to the sea bed authority and would have to divide their claim into two equal parts, one to be mined by the 'enterprise', the proceeds of which would be distributed under the common heritage principle. These terms were unacceptable to many Western nations, in particular the US, UK and the Federal Republic of Germany. These countries, along with France, Japan and Italy, have all formulated their own national legislation governing exploitation in international waters and have established a 'Reciprocating States Agreement', ostensibly as an interim measure, to prevent overlapping claims by member states.

Given the ambiguous situation pertaining to deep-sea mineral rights in international waters it is very unlikely that any operator will risk the huge amounts of capital necessary to develop a system for operation outside national jurisdiction. However, MPS deposits do sometimes occur within EEZs, and in that case there are no obvious legal reasons why mining should not take place, within a framework of national legislation formulated by the relevant coastal state. Such a process is in the early stages of development in the US, where policy-makers, scientists and government representatives are working to establish a legal and environmental infrastructure for the proposed leasing of the Gorda ridge for MPS exploration/exploitation (McMurray 1986).

The final non-technological factor affecting the feasibility of any proposed MPS mining operation is environmental considerations. The MMS produced a Draft Environmental Impact Statement which attempted to assess the environmental repercussions of a hypothetical MPS mining operation on the Gorda ridge (Zippin *et al.* 1983). The predicted effects ranged from drastic for the associated vent fauna to minimal for primary productivity. Many of the report's conclusions were necessarily vague owing to the high degree of supposition it incorporated; the criticism it provoked eventually led to the postponement of the leasing. The Gorda Ridge Task Force was then established, charged with collecting more relevant data, such as detailed knowledge of the mode of extraction likely to be employed.

The Red Sea Project has been subject to strict environmental monitoring since its inception. A sediment plume is produced as unwanted material is returned to the sea after primary concentration of the muds. It was originally discharged at 400 m depth but this had to be increased to 1000 m when it was found that the plume was being dissipated over a large area, giving rise to the danger of toxic effects from soluble components, and the possibility of particulate matter cutting out essential sunlight and/or interfering with the feeding of siphonophores (Thiel *et al.* 1986). The increased depth of discharge appears to have alleviated the problem, but it is in the nature of many environmental problems that adverse effects do not become apparent for years, and so long-term detailed monitoring of such novel operations is essential.

CONCLUSIONS

Knowledge of the distribution and average composition of MPS deposits is at a rudimentary stage. Continued basic research is necessary to remedy this situation, and also to add to the already considerable benefits from vent-related work to the sciences of geology, biology, and oceanography. The research and technological requirements necessary for the evaluation of the resource potential of MPS deposits are within the scope of current capabilities. However, it is unlikely that government funding will be forthcoming for any more than basic underpinning research; further development will be dependent on industrial involvement. Also, to be viable, such an evaluation must also consider non-technological factors pertinent to any proposed mining operation. Such an operation will only become a realistic possibility when world metal markets are more buoyant, and would then be feasible only under a secure legal regime formulated with MPS specifically in mind and paying due regard to environmental considerations.

REFERENCES

Carnahan, T.G. (1985) 'Treatment of metalliferous sulphides for market', *Marine Technology Society Journal* 19: 62–4.

Crawford, A.M., Hollingshead, S.C., and Scott, S.D. (1984) 'Geotechnical engineering properties of deep-ocean polymetallic sulphides from 21°N EPR', *Marine Mining* 4: 337–533.

Edmond, J.M. (1983) 'Chemistry of the 350° hot springs on the crests of the EPR at 21°N', *Journal of Geochemical Exploration* 19: 491–2.

Edmond, J.M., Agterberg, F.P., Bäcker, H., Delaney, J.R., Diehl, P., Dobson, M.R., Francis, T.J.G., Koski, R.A., Monteiro, J.H., Moorby, S.A., Oudin, E., Scott, S.D. and Speiss, F.N. (1987) 'Report of the Group on Marine Sulphides', in P.G. Teleki, M.R. Dobson, J.R. Moore and U. von Stackelberg *Marine Minerals: Advances in Research and Resource Assessment*, NATO

ASI series, Dordrecht: Reidel, pp. 29–37.

Harvey, W.W. (1986) 'Polymetallic Sulphides and Oxides of the US EEZ: Metallurgical Extraction and Related Aspects of Possible Future Development', draft report, Arlington Technical Services, USA.

Haymon, R.M. (1983) 'Growth history of hydrothermal black smoker chimneys', *Nature*, 301: 695–8.

McMurray, G. (1986) 'The Gorda Ridge technical task force: a co-operative federal–state approach to offshore mining issues', *Marine Mining* 5: 467–75.

Mustafa, Z., Narwab, Z., Horn, R., and Lelann, F. (1984) 'Economic Interest of Hydrothermal Deposits', Proceedings of the Second International GERMINAL Seminar, Brest, pp. 509–39.

Rona, P.A. (1983) 'Exploration for hydrothermal mineral deposits at sea floor spreading centres', *Marine Mining* 4: 7–37.

Rona, P.A., Klinkhammer, G., Nelson, T.A., Trefry, J.H., Elderfield, H. (1986) 'Black smokers, massive sulphides and vent biota at the mid-Atlantic ridge', *Nature* 321: 33–7.

Sawyer, D.L., Shyres, G.A., Sjoberg, J.J., and Carnahan, T.G. (1983) 'Cl^2–0^2 Leaching of Massive Sulphide Samples from the Southern Juan de Fuca Ridge, N. Pacific Ocean', Technology Progress Report 122, Bureau of Mines Extractive Metallurgy Program, Spokane, Wash.: US Bureau of Mines.

Scott, S.D. (1984) 'Polymetallic Sulphide Deposits of the Ocean Floor: Current Knowledge and Future Prospects', presentation on behalf of Citizens for Ocean Law, Washington, D.C., for the United Nations Preparatory Commission for Implementing the Law of the Sea, Kingston, Jamaica.

Siapno, W.D. (1987) '1986 Vinogradov expedition', *Marine Mining* 6: 223–9.

Sillitoe, R.H. (1982) 'Extensional habitats of rhyolite-hosted massive sulphide deposits', *Geology* 10: 403–7.

Thiel, H., Weikert, H., and Karbe, L. (1986) 'Risk assessment for mining metalliferous muds in the Red Sea', *Ambio* 15: 34–41.

Uyeda, S. (1987) 'Active hydrothermal mounds in the Okinawa back-arc trough', *EOS* 68 (36): 73.

Yates, J. (1988) 'A Comparative Study of Mining Technologies for the Exploitation of Marine Polymetallic Sulphides', Marine Resources Project, PREST, University of Manchester.

Zippin, J.P., Anderson, C.M., Baer, M., Beauchamp, B., Beittel, R., Climato, J., Cruickshank, M., Freidman, E., Hogue, E., Lane, J.S., Lewis, J.E. Sullivan, T.F., Tremont, J., and Turner, M.A.(1983) 'Proposed Polymetallic Sulphide Minerals Leasing Offer', draft environmental impact statement, Reston: Minerals Management Service, US Department of the Interior.

ACKNOWLEDGEMENTS

This study formed part of the 1987/7 programme of the Marine Resources Project (the marine division of the Programme of Policy Research in Engineering, Science and Technology), University of Manchester, and was funded jointly by the Department of Trade and Industry and the Marine Technology Directorate.

Part IV

Developments in general management: strategic decision-making

Part IV

Developments in general management strategic decision-making

9 Technology strategy in ocean management

Adrian F. Richards

Strategic management is a method of long-range planning that has gained wide acceptance in business. Technology strategy is the part of strategic management that couples the identification of new technologies and markets with business development. Businesses include profit-making firms as well as service organizations. Concepts of decision analysis, strategic management and technology strategy are described in the first part of the chapter. A case study is presented in the second part to show how the concepts were applied to a European organization concerned with assessing future research areas in deep-sea science and technology. The third part of the chapter suggests how managers can utilize information technology in their technology strategy plans.

INTRODUCTION

This chapter has three main purposes: (1) to describe briefly general concepts of decision analysis, strategic management and technology strategy, (2) to show application using a case study, and (3) to suggest that information technology is a valuable part of technology strategy. Strategic business principles and methods are essential to modern management. This chapter is concerned with the interrelationships of technology and management and suggests that managers in marine science and technology, or ocean engineering, can benefit from a better understanding of what technology strategy can do for them. In the context of this chapter, technology is used in the business sense of the word to connote science and technology rather than technology alone.

The training of an engineer implicitly involves planning and design, risk analysis, optimization and cost–benefit analysis, modelling theory and practice, and technology assessment and forecasting. These are subjects that broadly are as useful to management as to the profession of engineering. Most of these subjects can be grouped together under the rubric of decision analysis, which is the selection of an action from a number of alternatives. Decision analysis involves (1) obtaining information to structure the problem, (2) defining goals or operating principles, (3) considering applicable constraints and (4) evaluating and comparing alternatives to arrive at a decision.

Some time back the business world, recognizing the utility of decision analysis, adopted it under the name of corporate planning or, more recently, strategic management. Business rather than engineering terminology will be used in this chapter. But the methods are essentially the same even if the vocabulary is different.

A fundamental rule of strategy is to persuade your competitors not to invest in those products, markets and services where you expect to invest the most (Henderson 1984). Strategic management starts by asking three questions. What business should we be in? How are we now positioned to compete in that business relative to others? What strengths, weaknesses and range of resources do we possess relative to those of our present and potential competitors if we were to develop new businesses and new markets? There are some differences in the business world between strategic management and strategic planning, but they are considered insignificant for the purpose of this chapter.

Technology strategy is the management of technology for sustainable strategic advantage (Harris *et al.* 1984). It may also be considered to be a firm's approach to the development and use of technology (Porter 1985). The author includes under technology strategy the assessment of research and development effectiveness, efficiency and market integration.

Strategic management and technology strategy are essential for both profit-making (industry) and non-profit-making (foundations, government and university) organizations, which are considered together in this chapter. Both types of business are concerned with obtaining money, defining areas of activity to be financed and matching resources and know-how to achieve goals and objectives.

A short discussion of the use of the concepts of strategic management and technology strategy is given in the next part of this chapter. A case study will then illustrate the application of some of the concepts. Finally, the role of information technology in strategic management will be briefly discussed.

Table 9.1 Steps in strategic planning

1	Problem statement (determination of objectives and targets)
2	Forecasting to determine deficiencies
3	Appraisals of strengths, weaknesses, opportunities and threats (SWOT)
4	Description and evaluation of alternative strategies
5	Selection of the best strategy
6	Formulation of an action plan and a budget
7	Continuous monitoring or reviewing (feedback)

Source: modified from Argenti (1980).

CONCEPTS

Strategic management

Strategic planning, portfolio and competitor analysis, competitive positioning and scenario planning are components of strategic management. A simplified list of strategic planning steps is presented in Table 9.1. As previously indicated, these steps are approximately analogous to the elements of decision analysis. In Table 9.1 SWOT (an acronym for strengths, weaknesses, opportunities and threats) is a particularly useful concept, having wide-ranging applicability in the critical assessment of an organization relative to its competitors. Hussey (1985) provides additional details. The other components will not be separately discussed because of space limitations. Good descriptions have been given by Harris *et al.* (1984) for technology portfolio development, by Porter (1980) for competitive strategy, and by Huss and Honton (1987) for methods of scenario planning. In addition, Drucker (1985) and McNamee (1985) provide an introduction to basic knowledge, and Mason (1986) lists and discusses six phases of strategic analysis.

Technology strategy

There is less professional agreement on the formalized components of technology strategy. This is probably because the subject is relatively new and has not yet been codified, in addition to the fact that the complexity of technology usually requires somewhat different strategic methods to solve different problems. Table 9.2 presents one approach. In this table, technology portfolio development is a method for the identification and systematic analysis of key corporate technology alternatives and for the establishment of priorities in technology. In the case study that follows a quite different method of technology strategy, compared to that given in Table 9.2, was adopted.

Table 9.2   Components of technology strategy

1	Technology situation assessment (the internal and external scan of the technology environment beyond the limits of the traditional business portfolio)
2	Technology portfolio development
3	Technology and corporate strategy integration
4	Technology investment priorities

Source: after Harris *et al.* (1984)

CASE STUDY

Strategic plan

The case study selected to illustrate technology strategy in ocean management is a simplified one taken from an actual study (Richards 1986: hereafter called the report) together with additional commentary by the author. The first part of the plan follows the steps presented in Table 9.1.

Problem statement

The Joint Research Centre (JRC), Ispra Establishment, of the Commission of the European Communities, has had a long involvement with the study of the feasibility and safety of the disposal of heat-generating wastes into deep oceanic geological formations. Out of these studies have come many scientific results as well as a number of new technologies. The JRC has both developed new technologies and contracted with third parties for development. Several years ago, with the realization that the programme might be terminating before the end of the decade, it is assumed that management decided to investigate how its deep-sea research thrust and new technologies could be usefully and effectively applied to other deep-sea research areas.

The problem statement, or the objective, was: 'to analyse existing and future deep-sea problems and relevant technologies by identifying and overviewing significant deep-sea problems in the period 1984–94, with an identification of the new technologies that may be used to investigate (the) problems' (report). This information was expected to be helpful to the JRC in identifying new areas of science and technology for the purpose of reprogramming future activities.

Table 9.3 Brief SWOT analysis

Strengths	Weaknesses
Science	Science
Geochemistry	Benthic biology
Technology	Geology and geophysics
Modelling	Technology
Numerical analysis	Instrumentation
	Geotechnology

Opportunities	Threats
Previously stated	Not considered

Forecast

The JRC project team forecast that if the existing programme ended there would be an opportunity to use scientific capacity and technological products to enter new research areas and possibly to develop new markets for its services. As the technologies were highly original, there was probably limited concern that they would be taken over by competitors. Market opportunities might be limited because studies of the sea floor in water depths of 4000 m to 6000 m were likely to represent a restricted and highly specialized area for development. On the other hand, JRC leadership in ocean basin studies might provide a driving force or the stimulation that would encourage national efforts in the deep sea by European scientists and engineers and ultimately result in increased funding by the European Commission to the JRC.

SWOT

Table 9.3 lists the author-perceived strengths and weaknesses of marine and marine-related science and technology research and development capabilities in the JRC, which formed part of a SWOT analysis, although they were not identified as such in the report. Some areas indicated as weak in Table 9.3 were actually classified as weak to moderate or moderate in the report; these have been arbitrarily placed under weaknesses here.

Alternative strategies

This part of the plan was not considered in the report. Alternative strategies that might have been contemplated would include continuing business as usual or the eventual elimination of the group of people responsible for the programme. Neither could be considered a tenable strategy under the circumstances.

Action plan

This part of the strategic plan is discussed under technology strategy, which follows shortly. The preparation of a budget would be the next important step, but it is not applicable to this chapter.

Monitoring of reviewing

This part also is not applicable.

Technology strategy

The second part of the plan was to take action to solve the assigned problem. In the first step, the literature was reviewed, using automated (data base) searches and personal networking among knowledgeable experts. This effort identified a significant number of published summaries of problems and recommendations by scientific committees for future research.

The second step was to read and evaluate the literature obtained in step one to select those publications considered to be important and relevant. Information given in Table 9.3 was helpful in this evaluation.

The third step was to select from the publications passing the screen of step two those topics or subjects that were considered worthy of being included in the report. Again, for this purpose, information contained in Table 9.3 was used.

The fourth step was to synthesize information from the results of step three that would be presented to the JRC as conclusions and recommendations in the report.

Table 9.4 very briefly summarizes one result from step three for each of the science and technology research areas that were proposed. In this table, references to the source of information are not included for the sake of brevity; they were given in the report. The wording in the table has been taken from the report.

The principal conclusion from step four, reflected in the individual

Table 9.4 Selected marine science and technology research areas 1984–94

Benthic biology
What are the assessments of the metabolic requirements of benthic animals, including the relative importance of benthic ingestion of organic matter in sediments?

Geochemistry
Do the sedimentary surfaces influence the chemical potential of the diffusing solute so that the diffusion flux is not merely a function of concentration gradients? Do reactions between pore water and solid phases disrupt the flux resulting from concentration gradients?

Geology and geophysics
What are the processes that lead to the post-depositional alteration of the organic composition and clay mineralogy of marine sediments, to their lithification and cementation, and to the development of their acoustic and geotechnical characteristics? What factors govern the rates of these processes?

Geotechnology
Pore pressure and state of stress, particularly *in situ* and in areas of concern.

Technology
Mapping the ocean floor using autonomous vehicles. Unmanned, instrumented underwater laboratories, based on free-fall and return-to-the ship capability and permanently emplaced instrumented penetrators, for short and long-term experiments.

---

research areas given in Table 9.4, was that there was one significant problem common to all the science and technology areas originating out of step three that was considered to be highly relevant to the interests of the JRC. This interdisciplinary and multi-disciplinary problem concerned all scientific and engineering properties and processes from the surface of the sea floor to some depth in the sediments of the ocean basins. It included the distribution and origin or geochemical and geological fluxes, the geochemical and physical processes of diagenesis and the entire sedimentary hydrological regime. In short, this research appeared to be suitable for the JRC, and it was an important conclusion and recommendation arising out of step four. It was one that could maximize the new deep-sea technologies that had been developed and it also had the potential to open up new markets.

INFORMATION TECHNOLOGY IN TECHNOLOGY STRATEGY

The report cited in the case study was prepared in the mid-1980s, and the simple techniques that were used have been partly superseded by more sophisticated methods originating from the field of information technology. This field has become increasingly important because timely and complete information is a critical aspect of most technology strategies. Burrows (1986) describes some of the new information services for planners. They include using the microcomputer as a work station and information centre, on-line data bases, networking and electronic mail, and expert and knowledge-based systems for planners.

The fourth-generation computer language (Martin 1984), a high-level language that can automatically generate program instruction lines from simple input commands by users who are not programming experts, is rapidly evolving to the fifth generation. This includes the addition of parallel computer processing, user-friendly artificial intelligence software and many other possibilities that are described by Whaley and Burrows (1987). Höhn (1986) discusses business aspects of the rapidly emerging field of information technology and presents a number of methods that have applicability to modern management. The information given can be related almost equally well to academic, governmental and industrial ocean businesses.

An integrated information system is being developed by the author's company that utilizes a variety of information sources (Figure 9.1) to obtain scientific, technological and business information. Although the emphasis is on ocean-related subjects, the system has applicability to all fields.

Information technology coupled with technology strategy represents a powerful combination for ocean management or for the management of any other field. There is evolving a new type of technology strategist, one who utilizes or provides brokered information, has a knowledge of the relevant aspects of information technology and also utilizes an electronic office for information acquisition and communication with clients.

CONCLUSIONS

Strategic management and technology strategy are management techniques that are very relevant to the field of modern management. They appear not well known to, or used by, many ocean managers. Increased use will be beneficial to the profession.

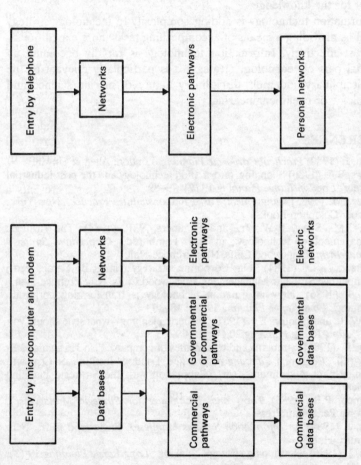

Figure 9.1 Generalized diagram showing how to acquire computer-assisted information

A strategic plan can help to minimize the adverse affects of 'environmental turbulence' or rapid technological and economic change. This can lead to increased profits for industry. A strategic plan can also lead to increased productivity and better utilization of human and material resources by service organizations, which include government laboratories and universities. With such a plan, research and development can be made more efficient, effective and integrated with the market for the knowledge.

Information technology is adding complexity to technology strategy as well as providing a means of accomplishing tasks more expeditiously and cost-effectively. Information technology is rapidly becoming an essential part of technology strategy. It is particularly relevant to the interdisciplinary and multi-disciplinary nature of marine science and technology, or ocean engineering.

REFERENCES

Argenti, J. (1980) *Practical Corporate Planning*, London: Allen & Unwin.

Burrows, B.C. (1986) 'Planning, information technology and the post-industrial society', *London Range Planning* 19 (2): 79–89.

Drucker, P.E. (1985) *Management: Tasks, Responsibilities, Practices*, New York: Harper Colophon Books.

Harris, J.M., Shaw, R.W., Jr., and Sommers, W.P. (1984) 'The strategic management of technology', in R.B. Lamb (ed.) *Competitive Strategic Management*, Englewood Cliffs, N.J.: Prentice-Hall.

Henderson, B.D. (1984) 'On Corporate strategy', in R.B. Lamb, (ed.) Competitive Strategic Management, Englewood Cliffs, N.J.: Prentice-Hall.

Höhn, S. (1986) 'How information technology is transforming corporate planning'. *Long Range Planning* 19 (4): 18–30.

Huss, W.R., and Honton, E.J. (1987) 'Scenario planning – what style should you use?' *Long Range Planning* 20 (4): 21–29.

Hussey, D. (1985) 'Strengths and weaknesses of companies', in J. Fawn and B. Cox (ed.) *Corporate Planning in Practice*, London: Institute of Cost and Management Accountants, (in collaboration with the Strategic Planning Society).

McNamee, P.B. (1985) *Tools and Techniques for Strategic Management*, Oxford: Pergamon Press.

Martin, J. (1984) *An Information Systems Manifesto*, Englewood Cliffs, N.J.: Prentice-Hall.

Mason, J. (1986) 'Developing strategic thinking', *Long Range Planning* 19 (3): 72–80.

Porter, M.E. (1980) *Competitive Strategy: Techniques for Analysing Industries and Competitors*, London: Collier Macmillan.

Porter, M.E. (1985) *Competitive Advantage: Creating and Sustaining Superior Performance*, New York: Free Press.

Richards, A.F. (1986) 'Review of some potential deep sea research areas over the decade 1984–1994', in *Study of the Feasibility and Safety of the Disposal of Heat Generating Wastes into Deep Oceanic Geological Formations*, Part 2, *Geochemical and Geotechnical Studies*, JRC Report Series, SP. 107, c2. 86, 17, GEC, Directorate General for Science Research and Development, Joint Research Centre, Ispra, Italy.

Richards, A.F. and E.A. (1988) 'Computer integrated information model for knowledge acquisition', in B.A. Schrefler and O.C. Zienkiewicz (eds) *Computer Modelling in Ocean Engineering*, Rotterdam: Balkema.

Whaley, R., and Burrows, B. (1987) 'How will technology impact your business?' *Long Range Planning* 20 (5): 109–17.

ACKNOWLEDGEMENTS

I acknowledge with appreciation the help given to me by my partner, E.A. Richards, in preparing this chapter. Drs Ing, C.N. van Bergen Henegouw, SOZ, The Hague, Ir, J.C. van Ingen, IHE, Delft, and Ir, J. deRuiter, Nieuwenkoop, reviewed the first version and offered helpful comments. Some of the information in the chapter was taken from my course notes for 'Introduction to Technology' taught at Lehigh University. The case study cited was performed when I was Vice-president for Research and Development in Fugro Geotechnical Engineers, Leidschendam. The opinions presented are solely mine; they are not to be construed as the opinions of Fugro or the Joint Research Centre in Ispra.

10 The Canadian Coastal Sediment Study

Overview of a major scientific and engineering research project

D.A. Huntley

The Canadian Coastal Sediment Study was a four-year project, running from 1982, designed to improve predictions of sand transport rates on beaches. The study had an overall budget of $1.3 million and was funded by four departments of the Canadian federal government.

The study had three co-ordinated parts; field measurements, the application of predictive techniques and the development of new instruments. The field programme, which occupied sites in the Gulf of St Lawrence in 1983 and 1984, was designed primarily to provide data with which to evaluate predictive formulae. Most measurements of this kind were carried out under contract by private companies. In addition several university groups took part in the field programme, carrying out more fundamental research but making extensive use of the data available from contractors. The interplay between these 'engineering' and 'scientific' aspects of the programme provided one of the most difficult yet interesting aspects of the study.

Evaluation of predictive formulae involved hindcasting offshore wave climate, refracting and shoaling the waves inshore and applying sediment transport equations. This part of the study confirmed the difficulty of obtaining even order-of-magnitude predictions, but led to identification of some specific areas where improvements were needed, particularly in the parameterization of energy loss as waves propagate from deep water and of bottom friction in the nearshore zone.

Instrument development concentrated on the new techniques for measuring rapidly varying suspended sand concentration. An optical technique used existing sensors and proved very valuable. An acoustic technique required the development of a new sensor, the Acoustic Suspended Sediment Profiler (ASSP). Unfortunately design and construction faults prevented this instrument from working as expected.

Nevertheless the experience gained has sparked continuing work on the development of a field sensor.

INTRODUCTION

The Canadian Coastal Sediment Study dominated Canadian coastal research, and the research of a number of non-Canadians associated with the study, for several years centred on 1983 and 1984. As a major participant in the study from its inception (though not directly involved in its administration) I was aware of many of the major benefits and also some of the problems associated with it. The purpose of this chapter is to give a personal retrospective view of the study. The study aimed to bring together scientists and engineers, government departments, government laboratories, universities and consultants to work towards a common goal. I hope that a discussion of the project, albeit a personal, undoubtedly biased view, will be of some interest to those planning similar collaboration work.

HISTORICAL BACKGROUND

The National Research Council of Canada (NROC) seeks to co-ordinate and inspire research in particular fields of interest through a number of associate committees. These committees have access to funds to allow for regular meetings and to provide seed money for activities such as conferences, workshops, newsletters, etc.

Recognizing a need to encourage research on shoreline erosion, harbour siltation and other problems connected with nearshore sediment movement, NROC established the Associate Committee for Research on Shoreline Erosion and Sedimentation (ACROSES) in the late 1970s, with members from government departments, government research institutes, universities and consulting companies. ACROSES quickly identified as one of its main functions the organization of workshops to clarify the needs in their broad research area. The first ACROSES workshop was held in October 1979, with the catchy title 'Workshop on Instrumentation for Currents and Sediments in the Nearshore Zone'. A number of recommendations were produced (Huntley 1979), the most significant of which were, first, that fast-response sediment monitors should be more widely used and tested, with development of new sensors to increase the scope and reliability of such instruments, and, second, that: 'more collaboration between different institutes, in particular between government institutions and universities, and between engineers and geologists, should be encouraged by ACROSES and other organisations'.

ACROSES took these recommendations to heart and, working through an energetic and diplomatic chairman, persuaded three of the four government departments involved in coastal management (Energy, Mines and Resources, Fisheries and Oceans, Public Works), in addition to the NROC, to fund a co-operative field study over three years, beginning on 1 April 1982. The project was named the Canadian Coastal Sediment Study, shortened to C^2S^2, thus claiming the world's first acronym to use letters raised to a power. Those without superscripting on their word processors were forced to call the project C2S2, and this was the label given to the series of reports published by NRCC as the study proceeded. The unifying theme of the project was to evaluate the techniques in common use by coastal engineers for estimating waves, currents and associated sediment movement at coastal sites. This evaluation was to be achieved in three main thrusts, acting concurrently:

1 *Field Measurements*. As many different and complementary measurements of coastal conditions as possible were to be made at a chosen field site, occupied for about two months on two consecutive years. Special emphasis was placed on the measurement of net alongshore sand transport, probably the most important physical parameter in coastal management.
2 *Prediction equations*. Existing prediction equations for waves, currents and sediment movement were to be applied to the chosen field site, and compared with measurements made in the field programme.
3 *New instrumentation*. Existing fast-response sediment sensors were to be tested, and new sensors developed, in response to the first recommendation of the ACROSES 1979 workshop. New instruments were to be ready for field testing during the second year of the field programme.

ORGANIZATION

A steering committee was set up by ACROSES to run the C^2S^2 programme. Members of the committee were drawn from the sponsoring government departments, from universities, from consulting/industrial companies and from the provinces in which the fieldwork was to be carried out.

The budget for the study was $330,000 per year for three years, plus two full-time positions within NRCC (the steering committee chairman and a full-time study manager) and some peopletime from the Atlantic Geoscience Centre of the Geological Survey of Canada, particularly for site surveys and field logistic support. One major restriction on the funding was that much of it was expected to be spent with the private sector, by contracting out. This had some advantages but also caused a number of problems, particularly in stretching the budget and in causing some loss of control over the work which was done. More will be said about this aspect of the study later.

Support for university research generally did not fall into the category of allowable contracts, though a few groups received funds, for example to build equipment (University of Toronto) or provide measurements (University of Quebec). However, university participation was encouraged by the provision of logistic support and operating funds for participation in the field programmes. There was also the prospect of data from a wide array of sensors which would be available as a result of the study. Limited operating funds were also used to allow the participation of non-Canadian researchers, including people from the University of Miami in Florida, the University of Southern California and the University of East Anglia.

At the end of the three-year period a further year was granted. This allowed completion of data quality control and presentation, so that data would be available in an intelligible form to any interested party, and also gave time to finish work on the comparisons between prediction equations and data.

With the inevitable budget overruns and the extra year, the final total cost of the study was $1.3 million. The proportions of the budget spent on different aspects are shown in Table 10.1. There will be some discussion of this table later.

Table 10.1 Spending of the C^2S^2 programme

Aspect	$	%
Equipment purchases and maintenance	280 000	22
The acoustic suspended sediment profiler	150 000	12
Evaluation of prediction procedures	130 000	10
Data acquisition and analysis	560 000	43
Support for basic research	110 000	8
Reporting and communication	70 000	5

THE STUDY

Fieldwork

The problem of selecting a suitable site for the fieldwork surfaced early in the meetings of the steering committee, with a split between engineers and scientists. Scientists were looking for a beach with simple topography, abundant sand and a clear, uniform exposure, on which they could test simple models of nearshore processes. The engineers, on the other hand, wanted a site which presented a real engineering problem, where harbour siltation or coastal erosion would provide a test of predictive models under real conditions rather than the simplified conditions for which they were originally derived.

Unfortunately no compromise field site could be found, and the engineers won the first round by the selection of Pte Sapin, New Brunswick, as the site of fieldwork in October and November 1983. Pte Sapin, described in detail by Forbes (1987), is partially sheltered to the south by Prince Edward Island in the Gulf of St Lawrence and faces south-east, so that sediment movement is essentially unidirectional to the south-west. An offshore breakwater at the southern end of the beach is designed as a sand trap, preventing sand entering the adjacent harbour mouth. The trap was dredged at the beginning of the field study and subsequent accumulation was used to assess net alongshore sand transport. The major and unexpected disadvantage of this site, affecting both scientific and engineering interpretation of the results, was the very limited sand cover. Despite an apparently abundant supply of sand on the beach face, the beach sand wedge was very narrow, flattening to a bedrock platform in a water depth of only 2 m. An additional sand body offshore possibly interacted with the littoral sand and added uncertainty to the estimates of sand transport. With the narrow littoral sand wedge it was impossible to mount instruments as densely as we would have liked, and the resolution of the direct measurements of currents and sand movement was therefore limited. Nevertheless a variety of sensors were deployed at the site. These sensors included a pitch-and-roll buoy in 15 m depth to measure the deeper water directional wave spectrum (deployment and data analysis by the federal government Department of the Environment), a current and pressure sensor system in 6 m depth to measure the inshore directional wave spectrum (a contract with Woods Hole Oceanographic Institution), a cross-shore array of two-component electromagnetic current meters and pressure sensors, and several instruments to measure suspended sediment concentration (mainly purchased by C^2S^2 and deployed and operated under contract).

In addition alongshore sand transport was estimated, using radioactive sand dispersal (a contract with University of Quebec at Rimouski).

Up to six groups of basic researchers were also active at the beach site, studying geology, surf-zone hydrodynamics, bed forms, Lagrangian drift and remote sensing techniques. Despite its limitations, the large data set resulting from this fieldwork has proved vital to the evaluation of predictor equations, and is a continuing source for scientific research.

This first year of fieldwork was very expensive, and it became clear that only a scaled down version would be possible in 1984. With work on the prediction equations still in progress, there did not appear to be any compelling reason to return to Pte Sapin, so the steering committee chose a 'scientific' site, Stanhope Lane, Prince Edward Island, for the second field experiment in October and November 1984. This beach is long and relatively straight, with at least three offshore, shore-parallel sandbars (Forbes 1987). Though sand is abundant, a gravel base is evident in the troughs between the bars. The relative lack of alongshore variability and the existence of bars attracted the scientists, but the exposure of the site to waves from a variety of directions caused sand transport to both east and west, and there was no sand trap to provide estimates of net transport rates. The site was primarily used, therefore, to conduct basic scientific studies and as a field test-bed for suspended sediment sensors.

The newly developed acoustic sediment profiler (to be discussed later) was first exposed to field conditions at Stanhope Lane and proved unequal to the challenge. However, another acoustic suspended sediment profiler brought up from the University of Miami, in collaboration with Dalhousie University, worked well. Other research groups continued their work on nearshore waves and currents, bed forms, bar movement, radioactive sand tracer movement and suspended sediment dynamics.

Prediction techniques

The work of applying existing prediction equations to both field sites was done by coastal engineering consultants under contract. Prediction of net alongshore sand transport was a four-stage process:

1 Offshore wave spectra were hindcast, using measured coastal winds.
2 A spectral wave refraction model was used to transform the deep-water spectra to a nearshore location.
3 Longshore currents generated in the surf zone were predicted either from measured or from predicted nearshore wave spectra.
4 Potential alongshore sand transport rates were then calculated either

direct from the inshore wave spectra or using the calculated alongshore currents.

At each stage predictions were compared with observations. Generally offshore wave predictions were in reasonable agreement with spectra measured with the pitch-and-roll buoy, though there was some uncertainty in estimating offshore winds from coastal stations. However, major problems arose at each of the other stages of prediction. Predicted nearshore wave heights were typically two or three times larger than those observed, unless a 'saturation' criterion, presumably representing white-capping, was imposed. When comparing measured alongshore currents with predictions it was obvious that the results were sensitive to the poorly defined sea bed friction. Finally, with so many uncertain input parameters, predictions of net alongshore sand transport, even for the same storm, ranged over more than two orders of magnitude (Fleming *et al.* 1986). One of the two consulting companies involved in this work concluded that:

> no consistent conclusions regarding sand transport based on the predictor equations could be drawn. Different procedures resulted in widely varying transport rates, and opposite net directions, depending on the choice of theory, data source and methodology for estimating the various input parameters to the theory.

(Baird *et al.* 1986)

Instrument development

Work on the assessment and development of sensors to measure suspended sediment concentrations centred on two types of instrument, one optical, the other acoustic. The decision to concentrate on these two sensors arose from a survey of sensors commissioned by the C^2S^2 steering committee at the beginning of the study (Huntley 1982). The optical backscatter sensor (OBS) had been developed at the University of Washington to measure rapidly varying sand concentrations at a number of fixed points. At Pte Sapin two groups of these sensors were on loan from the University of Washington, and they proved so successful that C^2S^2 purchased its own sensors for Stanhope Lane a year later. Although there are still some questions relating to data quality control and possible instrument saturation, the OBSs proved their worth during C^2S^2 and are continuing to provide unique and exciting data.

The sensor survey commissioned by C^2S^2 (Huntley 1982) had concluded that the most promising technique for fast-response estimation of suspended sand concentrations was through the measurement of acoustic backscatter. Accordingly the C^2S^2 steering committee set out to develop such an instrument, hoping to have it ready for field deployment in 1984. A workshop was held to discuss the specifications of such a sensor, with participants from the Netherlands, Britain and the United States as well as Canada. In common with similar sensors developed elsewhere, the sensor was to consist of an acoustic transducer, held about 1.5 m off the sea bed, sending a narrow acoustic beam vertically downwards. The acoustic return would be range-gated to provide estimates of backscattered intensity at different distances below the sensor, to the sea bed, with a vertical resolution of about 1 cm.

The detailed design and construction of the new sensor, to be named ASSP (acoustic suspended sediment profiler) were contracted out to a private electronics company. Unfortunately the ASSP has never functioned as was intended, owing to design and construction faults. After delivery of the ASSP and its complete failure at Stanhope Lane, NRCC undertook some redesign and, after two years, got it working as a laboratory sensor, but it is unlikely that it will ever be suitable for field measurements. The brightest aspect of this part of the C^2S^2 programme is that is has led on to the development of a true field sensor by two university groups, funded by the Natural Sciences and Engineering Research Council of Canada. The new instrument is already beginning to provide interesting field data.

THE C^2S^2 IN RETROSPECT

As stated earlier, the primary aim of C^2S^2 was to evaluate engineering techniques for the prediction of nearshore wave, current and sediment conditions. C^2S^2 can be considered a moderate success in addressing this aim. Comparisons between the predictions and the observations, particularly at Pte Sapin, and between different prediction schemes on both beach sites, have revealed with greater clarity the parlous state of the art! In fact the lack of agreement between the various methods of estimating transport is so great that it is difficult to identify areas where effort is needed in order to improve predictions. However, the steering committee, in its final recommendations (Bowen *et al.* 1986), focused on the following primary needs:

1 Accurate offshore winds, as input to the hindcasting procedure.
2 Further investigation of wave transformation from deep to shallow water, particularly to identify the causes of energy dissipation prior to depth-limited wave breaking at the shoreline.

3　Research, both laboratory and field, into relationships between physical bed forms, hydrodynamic roughness and friction, related to both alongshore currents and sand transport.

4　Continuation of the kind of research exemplified by C^2S^2 so as to develop more reliable models of nearshore sediment transport.

The major disappointment with C^2S^2 has been that it stopped at the point where the problems had been defined and was not allowed to continue in order to develop solutions to some of the problems. In retrospect perhaps the balance of the budget, shown in Table 10.1, was too heavily weighted towards acquiring and analysing field data (65 per cent, including the purchase of equipment), with too little (10 per cent) being used in the evaluation of prediction procedures. More use of existing expertise in government research laboratories and universities, through relaxing the conditions on contracting out, might have saved some expenditure on the field programme and left more money for a thorough analysis of why prediction schemes show such wide scatter. More timely results from the Pte Sapin predictions could also have provided useful input to the design of the Stanhope Lane experiment. Nevertheless the list of recommendations given above puts the ball firmly back in the scientists' court. With subsequent research council support some of these problems are being addressed.

On the measurement side, considerable strides have been made in evaluating instruments. Although neither of the C^2S^2 sites was ideal, the field measurements showed just how close we are to the direct measurement of net transport of suspended sand through a shore-normal transect across a beach. Given the uncertainty associated with indirect estimates of transport (- even the sand trap measurements at Pte Sapin were uncertain to within a factor of two, depending upon the assumed porosity of the accumulated sand and assumptions about the bypassing of the trap), direct measurements are certainly worth pursuing. The continuing development of acoustic instruments, in particular, is an encouraging outgrowth of C^2S^2. However, C^2S^2 made no headway in the development of sensors for the direct measurement of *bedload* transport, and left this as a recommendation for future work.

On the management side I think it is fair to say that the process of contracting out major aspects of the study did not always work well. Mistakes were probably made on both sides and the resulting problems highlight what is probably a lack of experience with joint projects between researchers and industry. The demise of the ASSP exemplifies some of the problems. Development of the ASSP began well, with full discussions to define specifications and with enthusiasm about the prospect of a functioning sensor. However, once the work was contracted out, the lack of close communication between the contractor and those defining the nature of the contract meant that the final product was unsatisfactory. The solution, if work is to be contracted out, is probably to ensure that the contract is meticulously specified in the first place, to ensure that there is no ambiguity in what is required, and then to ensure that regular communication with the contractor is maintained.

The C^2S^2 project certainly fulfilled its aim of bringing together industry, government and university researchers, by sponsoring the joint field programmes. The benefits were obvious. The university workers enjoyed the logistic support provided by the government and industry workers, and continue to benefit from the broad data sets gathered under contract. At the same time the industry representatives, charged with gathering data for the engineering evaluations, learned from the experience of the university researchers in field techniques. Even the various university and government research groups, naturally closer in their aspirations, gained a new understanding of each other's work and benefited greatly from the discussion of mutual problems.

To some degree the collaboration between basic research groups, begun at the C^2S^2 field sites, is continuing, with joint programmes going ahead using other sources of funding. In 1987 two major joint programmes received funding, one between Dalhousie University, Nova Scotia, and the Memorial University of Newfoundland to develop a new acoustic suspended sediment profiler, and one between Dalhousie University and the University of Toronto to conduct fieldwork to investigate, among other things, the nature of sea bed roughness and friction in the nearshore zone. Thus the spirit of C^2S^2 lives on in these further studies. However, the contact between the basic research groups and the engineering and industrial/consulting sectors has been short-lived. It was primarily the existence of the C^2S^2 funds which caused the collaboration to begin, and when the money ran out joint work ceased. The C^2S^2 steering committee final report (Bowen *et al.* 1986) lists, as one of its main recommendations, that 'A clearly defined lead agency is needed for Canadian coastal research, to bring scientists, engineers and consultants together, and to provide financial incentives and logistical support for coastal research.' Without such a lead agency, a role taken by NRCC for the C^2S^2 project, collaborative projects seem unlikely to get off the ground.

Finally, C^2S^2 published a series of twenty-four reports of its work during its lifetime, and a final report. Willis (1987) lists these reports, which are available free of charge from NRCC, Ottawa.

REFERENCES

Baird, W.F., Readshaw, J.S., and Sayao, O.J. (1986) 'Nearshore Sediment Transport Predictions, Stanhope Lane', PEI National Research Council of Canada, Canadian Coastal Sediment Study Report No. C2S2–21, Ottawa: NRCC.

Bowen, A.J., Chartrand, D.M., Daniel, P.E., Glodowski, C.W., Piper, D.J.W., Readhsaw, J.S., Thibault, J., and Willis, D.H. (1986) 'Canadian Coastal Sediment Study, Final Report of the Steering Committee', Hydraulics Laboratory Technical Report No. TR–HY013, Ottawa: National Research Council of Canada, Division of Mechanical Engineering.

Fleming, C.A., Pinchin, B.M. and Nairn, R.B. (1986) 'Evaluation of Coastal Sediment Transport Estimation Techniques, Phase 2, Comparison with Measured Data', Canadian Coastal Sediment Study Report No. C2S2–18, Ottawa: National Research Council of Canada.

Forbes, D.L. (1987) 'Shoreface sediment distribution and sand supply at C^2S^2 sites in the southern Gulf of St Lawrence', *Proc. Coastal Sediments '87*, New York: American Society of Civil Engineers, pp. 694–709.

Huntley, D.A. (1979) 'Summary and Recommendations', Proceedings of the Workshop on Instrumentation for Currents and Sediments in the Nearshore Zone, Ottawa: National Research Council of Canada Associate Committee for Research on Shoreline Erosion and Sedimentation, pp. 191–4.

Huntley, D.A. (1982) '*In situ* Sediment Monitoring Techniques: a Survey of the State of the Art in USA', Canadian Coastal Sediment Study Report No. C2S2–1 Ottawa: National Research Council of Canada.

Willis, D.H. (1987) 'The Canadian Coastal Sediment Study: an overview', *Proc. Coastal Sediments ' 87*, New York: American Society of Civil Engineers, pp. 682–93.

ACKNOWLEDGEMENTS

Much of the factual information in this chapter is drawn direct from the summary by Willis (1987). Thanks to all the many participants in the C^2S^2 programme for a thoroughly enjoyable, and productive, couple of years of collaboration.

11 The management of data-gathering programmes

K.R. Deeming

This chapter is concerned primarily with the data-gathering programmes in the private sector, but many of the points must apply equally to long-term national programmes. It is procedural rather than technical. Recent years have seen a reappraisal of the marine science strategy of both the private and the public sector, in response to changing political and market forces. In the public sector there is considerably more interest in long-term projects of an international nature; in the private sector there has been greater emphasis on offering multi-disciplinary skills to solve estuarine and coastal problems. Although their projects are spatially and temporally different, both sectors require expertise in data collection and numerical modelling. The chapter outlines how a clearly defined set of procedures and techniques for a data-gathering programme will assist its successful management. In the private sector the philosophies of quality assurance and quality control, whether imposed from outside or not, will increasingly play their part in this. Also briefly discussed is the important question of how data should be archived after they have been gathered and analysed.

INTRODUCTION

We have been using, and sometimes abusing, the ocean for thousands of years, with little or no understanding of the implications of our actions. We extract its minerals and sediments; we exploit its mammal and fish stocks; we use it as a sink for our varied wastes. It is only relatively recently that we have begun to perceive that, for us to continue with these activities to our long-term benefit, we must make proper use of the resources available, i.e. we must manage those resources.

This awareness of the need for a cohesive approach to marine resource management is well recognized by Dexter (1988) and

Shackleton (1988). In reviewing both the national and the international dimension they talk of the need for strategies and tactics; of the need for us to blend and co-ordinate the necessary skills and disciplines; and of the need to measure and predict the ocean, and hence manage better.

To manage well we need a matrix of skills – many of them not scientific. The socio-economic advantages and disadvantages of developing, or maintaining, a marine resource have a major influence on our management decisions. In consequence, lawyers, economists, architects, engineers, urban planners and politicians – to name a few – have important roles to play. The ecologists, oceanographers, surveyors, geologists and geophysicists form just a small part of the matrix.

Nevertheless, in order for us to manage any marine resource successfully we need first an understanding of the physical, chemical and biological processes which quantify and qualify the resource. Once even a partial understanding has been achieved we are in a far better position to make socio-economic decisions on the development and conservation of that resource. This applies equally to the coasts (outfalls, fish farming, barrages, land reclamation), the shelf seas (gravel extraction, hydrocarbon exploitation, fishing), or the deep seas (waste disposal, alternative energy, mineral mining).

Currently, there is a reappraisal of how best the applied marine sciences should fit into the 'scheme of things'. The House of Lords select committee report *Marine Science and Technology* (1985) has acted as a major catalyst. In response, and to maximize the benefits from limited government finance, the public sector has addressed itself to those programmes which will be most cost-effective, yet nationally beneficial. The outcome has been the Natural Environmental Research Council's recently published *Strategy for the Marine Sciences* (1987), in which five key projects have been identified for specific attention over the next two decades, with emphasis on understanding the ocean in time and space.

Likewise, the private sector has had to review its objectives and markets with the down-turn in activity in the offshore oil industry. It has been helped by the Department of Trade's initiative in formulating the Resources from the Sea Programme (RSP), which is designed to help British industry seize the commercial opportunities available from developing marine technology for the exploration and exploitation of the oceans (Senior 1987). Also acting as catalysts in this re-evaluation have been the Control of Pollution Act, Part II, and the various European Community directives on water quality and pollution. In consequence there is greater emphasis on offering a suite of multi-disciplinary skills which are integrated to solve estuarine and coastal problems world-wide.

All these events are forcing us to be more cost-effective, with greater emphasis on quality and accountability.

In both the private and public sectors there is the requirement to predict the ocean and the effects of man's actions on it. Only the scales – both temporal and spatial – may be different. NERC is planning for the year 2000 and looking at global oceanic and shelf-sea problems; the private sector is more involved in coastal and shelf-sea studies for resources which man is developing, or wishing to develop, over the next few years. In each, there is the need to gather data and use the data 'to predict the ocean'.

With the phenomenal growth in electronics over the last twenty years has come undreamed of data-gathering and analytical capabilities. Numerical models, of one form or another, are increasingly being used in marine studies. They can range from simple statistical models to sophisticated three-dimensional ones. Computing power is said to be increasing by a factor of ten every six years. Britain's long-term partici- pation as a nation in international collaborative projects, such as the World Ocean Circulation Experiment, which will use extremely powerful computer systems will enable us to predict world ocean circulations. and associated climatic conditions, not years but decades ahead (WOCE 1985). At this level the computers will process billions of measurements, gathered throughout the breadth and depth of the ocean, for calibration and validation purposes. Once these models have been developed we shall have the ability to forecast and plan for storms and droughts, and the knowledge to make full use of the ocean's resources.

On shorter time scales, i.e. years instead of decades, models are being developed to predict the hydrodynamics and water quality of shelf seas (Howarth 1988). The intense commercial and environmental pressures on these shelf seas necessitate multi-disciplinary data-gathering programmes for baseline studies and model validation. They are also at an international level, with the management, or mismanagement, of one country's marine resources having a direct effect on neighbouring states.

These shelf and deep-sea research programmes will require substantial public finance and government resources in ships, equipment and personnel. The objectives are long-term, enabling careful planning and management; for example, the planning for WOCE has already been proceeding for five years, with its own international planning office and full-time scientific director to organize world-wide data-gathering programmes using satellites, research ships, a tide-gauge network, surface and deep floats, and computers for ocean modelling. Likewise the intensive monitoring programme scheduled for the North Sea in

1988/9 lasted fifteen months and the subsequent analyses and modelling will last for five years.

Contrast this with the more immediate requirements of commerce and governments for specific resource or engineering developments in the estuarine and coastal zones. In these the objectives are short-term and the responses more tactical than strategic, although numerical models and data collection still form an integral part of the work. Planning time may, at the most, be only three or four months. Nevertheless the same care and attention which will go into the long-term projects are needed here.

Perhaps even more so, for at the tactical level, when projects are driven by tight commercial and legislative guidelines, there is a need for sound management by both the data gatherer and the client, who pays for it all. For too long we have been interested in the manipulation and analysis of the data, without always questioning whether the data are suitable or not in the first place. The honours have gone to the analyser and modeller, without our realizing that professional and effective management of the data-gathering programme is vital if we are to meet engineering objectives successfully (Deeming and Jonas 1986).

DATA-GATHERING PROGRAMMES

Qualified to manage?

There is no one discipline which is more competent than another to manage a data-gathering programme, particularly if it is multi-disciplinary. In the private sector there has been a tradition that data-gathering projects have been managed for the client by the consulting civil engineer and for the data gatherer by the hydrographic surveyor. It was not because their disciplines were any more relevant than the others; far from it in many cases. It was because they had management experience and professional status, whereas the other disciplines had not – the civil engineer by virtue of his role as designer and builder and the hydrographic surveyor, in many cases, because of his naval training.

This has now changed. In any major data-gathering programme the representatives of the client and the data gatherer will be drawn from the wide spectrum of marine disciplines. What they will both have is the ability to understand the broad scheme of things, and to co-ordinate and control accordingly. Their remit is to measure and analyse a range of variables, make sense out of them – using models or otherwise – and provide the engineer or marine resource manager with results and information which he can be confident will be 'fit for purpose'.

The Land Surveyors Division of the Royal Institution of Chartered Surveyors has now recognized this fact and is offering marine scientists, with certain postgraduate qualifications and suitable experience, the opportunity to apply for membership through a test of professional competence, thereby providing the professional recognition which has been unobtainable in any other chartered body. In time, as this begins to be appreciated in both the private and the public sectors, the letters ARICS or FRICS in marine science will be of value. They will raise the standards of the profession and eventually the standards of our data-gathering programmes.

Organization

The mechanics of initiating and conducting a marine data-gathering programme have been covered to some extent by Deeming (1988), Jensen (1988) and RICS (1983). The first paper considers general data-gathering programmes, with some emphasis on oceanographic and environmental work, whereas the second and third consider the institution and administration of hydrographic and geophysical surveys. In all three the same basic philosophies and concepts apply, and they provide guidelines on how to structure and conduct the work with an emphasis on good management practice by both the client and data gatherer. If a long-term series of geophysical or oceanographic measurements are undertaken it is worth reading Tabata (1985), who considers problems, unique to this type of work, of personnel, equipment, logistics, liaison and quality control. The more technical details of a multi-disciplinary study are better considered by Elliot *et al.* (1988), and many of the ideas and techniques can be extrapolated offshore to deeper water.

The chronological sequence of an environmental, geophysical, oceanographic or survey data-gathering programme in the private sector may be as follows. (Although it is written from a client's point of view, much of it applies equally to the data gatherer managing the work independently, in both the public and the private sector.)

1 Define the client's aims and objectives.
2 After evaluating all available reports and data, define the data-gathering programme and the end products required, taking into account the financial budgets, including contingencies.
3 Prepare a sound scope of work, a comprehensive set of technical specifications and a realistic schedule of prices.
4 Pre-qualify suitable data-gathering organizations and then involve the selected parties in competitive pricing.

5 Conduct a thorough tender evaluation, taking into account technical content and grasp of the project, and not least the total cost.

6 After a pre-award meeting, tie the project down with a suitable and comprehensive contract. At this stage involve the data gatherer in as much technical decision-making as possible.

7 Undertake the fieldwork, using the data gatherer's quality assurance system. Both the data gatherer and the client should apply quality control to equipment, personnel and data. It is imperative that a continuous dialogue is maintained throughout.

8 Analyse and validate the data with the appropriate checks and assessments of 'scientific reasonableness'.

9 Prepare reports and quality control documentation.

10 Use the data, in conjunction with the data gatherer, to meet the client's objectives.

11 Archive the data on proprietary or national data banks, or on referral data banks.

Scope of work

The purpose of the 'scope of work' is to define the requirements for a particular data-gathering programme, drawing on the technical specifications for reference. It is here that the specifications can be amended and deleted.

Topics which should be addressed include, but are not restricted to:

1 *Objectives of the project*: what the client, i.e. the engineer or planner, wants to achieve using the data, and how they relate to his overall objectives.

2 *Measurement requirements*: where and when the work is to be done, using certain generic types of equipment, including location control and moorings if required.

3 *Schedule of work*: the mobilization and demobilization phases, the duration of the measurements programme, and the dates for the submission of draft and final reports and any computations or numerical models.

4 *Checks and calibrations*: the methods required, bearing in mind the objectives and duration of the project, the duration of the mobilization and demobilization phases, and any financial constraints.

5 *Validation and reporting*: the types of analyses required and the methods of presentation after the validated data sets have been prepared.

6 *Additional requirements*: any further changes to the basic technical specifications.

Technical specifications

The specifications for each variable should aim to be flexible enough to cover most types of instrumentation, yet detailed enough to ensure fair tendering and sound scientific work by the data gatherer. They need to be relevant to the work. They provide the framework upon which the scope of work can be hung. Experience has shown that more detailed 'method' specifications are preferred to 'performance specifications' which create uncertainties and poor objectives.

Each specification should include some, or all, of the following, irrespective of what is being measured:

1 Required accuracies.
2 Instrumentation checks and calibrations.
3 Methods of deployment and recovery.
4 Operation and maintenance.
5 Quality control of data.
6 Scientific assessment.
7 Data presentation and reporting.
8 Data archiving.

Field specifications

Complementing these technical specifications for each variable should be a 'field specification' for ancillary equipment and services used in data-gathering programmes. This should state the minimum requirements for:

1 Survey vessels.
2 Position fixing systems.
3 Radio transmission of data.
4 Notices to maritime authorities.
5 Moorings.
6 Safety practices, onshore and offshore.

Quality assurance

All these procedures are part of QA, or quality assurance. The operational technique which implement some of them are part of QC, or quality control. Neither QA nor QC is the prerogative of the client or his representatives; far from it. In the offshore industries, it is possible that data-gathering programmes, particularly in surveying and geophysics, have suffered because the client has applied his own quality control too rigorously, thereby allowing the data gatherer to abrogate his responsibilities in this matter. This has been counterproductive for both parties: the client, or his representative, has been so involved in applying quality control to the data gatherer he may have neglected his own quality assurance; and the data gatherer, to some extent, has allowed his quality assurance to be dictated to him. In consequence, professional standards have not been improved to the benefit of all concerned.

There are many definitions of quality assurance but, when applied to a data-gathering programme, the following is suggested: 'the management skills, documented procedures, field procedures and analytical methods necessary to ensure that the work is "fit for purpose" and satisfies defined needs with a minimum of expenditure'.

Likewise quality control 'is implemented through the operational techniques that sustain and maintain the quality of the service'. It is applied by both client and data gatherer during the progress of the work and subsequent analysis and reporting, including any numerical models, as described in detail by Copeland (1987).

These represent an overall definition of a quality system which is defined in quality assurance manuals. These manuals document a company's system and allow it to be assessed against British Standard BS 5750 (BSI 1987). This can be done by the client, or by a recognized third party accreditation body such as Lloyd's or the British Standards Institution. In marine data-gathering programmes this requirement of accreditation will not go away; there is continuing momentum within the EEC and the British government for it to happen (Hill 1988). Although accreditation will be costly it is welcomed for the fact that it will raise professional standards and nurture the good companies at the expense of the bad. It should also be appreciated that the definitions of quality assurance and quality control outlined above could apply equally to marine data-gathering programmes in the public sector, and it is hoped that the experience of the private sector over the next few years will be of value to the universities and government institutes.

DATA BANKS OR DATA BASES

The archiving of the validated data is complementary to the whole programme. There is enormous benefit, both for the data gatherer and for other potential users in the future, for the data to be stored securely and correctly in some form or other. The national data banks in the UK

– such as the National Oceanographic Data Bank (NODB) – are under-resourced and under-financed. In consequence, sets of good data are being lost to the nation and future marine resource managers.

To manage and quality-control large data banks is expensive. A more cost-effective solution is for the many data gatherers around the country – in both the private and the public sectors – to provide solely information on the data to a central clearing office. If information is required at any time in the future on such diverse matters as trace metals in the Bristol Channel, surface winds over the Irish Sea, beach studies along Chesil Beach, current measurements off Scarborough, sediment cores in the Firth of Forth, etc., the central clearing office is approached, electronically or otherwise, to see who owns any data and how they can be accessed. The clearing office acts as a 'Yellow Pages', pointing the enquirer in the direction of the research worker, proprietary data bank or industrial client which owns the data. It is a 'referral data base'.

In the USA there is the National Environmental Data Referral Service (NEDRES), run by NOAA, which identifies the existence, location, characteristics and availability of environmental data (all environmental data, not just marine). NEDRES describes the data and directs the user to the holder of the data, but the data themselves are not available from NEDRES (Barton 1987).

This could be a powerful tool for our marine resource management programmes over the next two decades. At present our national data banks are facing great difficulties financially. If we continue to finance them at the present level they will not be successful. A national referral data base brings the whole problem back into the market place, encouraging owners to trade or sell their data, and, paradoxically, raising standards by encouraging marine scientists to use their professional judgement in evaluating the quality and usefulness of any data set.

CONCLUSIONS

1 In the last two years both the private and the public sector have reappraised their strategies and tactics in applied marine sciences. This development is overdue and welcome.
2 In the public sector there is greater emphasis on long-term projects of an international nature and in the private sector there is an increased appreciation that integrated skills are needed to assist in marine resource management programmes.
3 Although their projects are vastly different in scale and time, both sectors require skills in data collection and numerical modelling.
4 There is a greater need for effective management of the data-

gathering programmes, and the various managers required should be drawn from the spectrum of marine disciplines, with the emphasis on management and communication skills.

5   A sound scope of work and comprehensive set of specifications will greatly assist the management of a project. They form part of a quality assurance system which, in the private sector, is being defined increasingly by British Standard 5750. It is possible that the public sector could benefit from applying similar techniques to its own data management programmes.

6   The management of any gathered data includes its quality control and archiving. Quality control should remain with the data gatherer, as should the data. Only information on the data should be provided to a national 'referral data base'.

REFERENCES

Barton, G.S. (1987) 'NEDRES: an interactive computer tool for locating geophysical information', *EOS* 68 (19), 12 May.

British Standards Institution (1987) *British Standard 5750: Quality Systems*, London: British Standards Institution.

Copeland, G.J.M. (1987) 'The Quality of Numerical Models', symposium on 'Modelling the Offshore Environment', 1–2 April, London: Society for Underwater Technology (Graham & Trotman).

Deeming, K.R. (1988) 'The quality assurance of data gathering programmes in the coastal zone', *Hydrographic Journal* 47, 23–6.

Deeming, K.R., and Jonas, P.J.C. (1986) 'Coherent policy needed for UK offshore data collection', *Offshore Engineer*, February, 34–5.

Dexter, K. (1988) 'The National Dimension', Royal Institution of Chartered Surveyors, conference on 'Marine Resources: Commonality and Conflict', 23 March, London: RICS.

Elliot, R.C.A., Paggett, R.M. and Raynor, R.E. (1988) 'Environmental Studies for Outfall Design, Construction and Monitoring', symposium on 'Aspects of Sea Outfalls', 20 January Exeter: Institution of Water and Environmental Management.

Hill, C.E.J. (1988) 'QA in land surveying - BS5750 is coming', *Land and Minerals Surveying* 6, February.

House of Lords (1985) 'Marine Science and Technology', House of Lords Select Committee on Science and Technology, No. 47–I, December.

Howarth, M.J. (1988) 'The (NERC) North Sea Community Project, 1987–1992', *Oceanology '88* Advances in Underwater Technology, Ocean Science and Offshore Engineering 16, London: Society for Underwater Technology (Graham & Trotman 1988).

Jensen, M.H.B. (1988) 'Quality assurance in positioning seismic surveys', *Hydrographic Journal* 48, 25–36.

National Environment Research Council (1987) *The Challenge: NERC Strategy for Marine Sciences*, London: Natural Environment Research Council.

Royal Institution of Chartered Surveyors (1983) *Guidelines for the Preparation of Specifications in Offshore Surveying*, Surveyors' Publications, G.L. Haskins and W.J.M. Roberts, London: Royal Institution of Chartered Surveyors.

Senior, G. (1987) 'Resources from the Sea Programme – Preamble', *Journal of the Society for Underwater Technology* 13.

Shackleton, Lord (1988) 'The International Dimension', Royal Institution of Chartered Surveyors, conference on 'Marine Resources: Commonality and Conflict', 23 March, London: RICS.

Tabata, S. (1985) 'Specific problems in maintaining time series observations', *IOC Time Series of Ocean Measurements* 2, *1984*, IOC Tech. Ser. 30, Paris: UNESCO.

WOCE (1985) *WOCE–IPO Newsletter* 1, Wormley: IOS Deacon Laboratory.

Part V

Developments in general management: regional contexts

12 The integration of coastal and sea use management

J.E. Halliday and H.D. Smith

INTRODUCTION

Recent decades have witnessed a now well documented expansion and intensification in the use of the world's seas (see, e.g. Couper 1983). The movement has been paralleled, or perhaps more accurately exceeded, by an expansion and intensification in the use of coastal land and nearshore waters (see, e.g. Clark 1987). This is a process which is fuelled not only by developments occurring offshore, which tend to have ramifications for the use of coastal land and waters, but also by independent forces of change originating on land.

One consequence of this increasing inventory of coastal and oceanic resources is the growing requirement for their effective planning and management, or the effective planning and management of their areas of origin. Instead, however, a variety of factors have combined to produce a series of disparate systems of control. These are typically characterized either by uncoordinated, use-specific directives or by more comprehensive but spatially limited initiatives, relating to specific areas and originating most frequently in response to crisis.

A self-evident corollary is that oceanic initiatives are conceived in isolation from related developments in nearshore waters and coastal land. Nearshore waters are frequently administered and managed as distinct from their neighbouring maritime and coastal environments and the third component of the trilogy, the coastal land, forms a further exclusive managerial dimension.

This chapter looks first at the evolution of coastal and sea use management along these distinct paths. It describes the key characteristics of the systems that now prevail and examines some of the limitations inherent in the current divided order. It then argues for the necessity of achieving both an increasingly integrated approach to management and an overall strategy for the determination of managerial priorities. This is illustrated by our suggestions for a technical and general management approach. Finally we describe some key characteristics of the present system which militate against the realization of such an idealized scheme.

THE RISE OF COASTAL AND SEA USE MANAGEMENT

As suggested above, the rise of coastal and sea use management parallels the increasing incorporation of these areas into the world's resource base. This movement has several important but related dimensions. The first is a function of rising demand and reflects the increasing claims made upon traditional sources of supply. Aggregates, for instance, have long been taken either from the land or from the shore, where beach deposits have been seen as an easily accessible and readily sorted store. Increased reaction to the environmental impact on land and increased evidence of the results of removing relict beach deposits have, however, diverted attention to offshore reserves. By the beginning of this decade, for instance, nearly one-fifth of the annual production of sand and gravel in Britain came from the sea bed and continental shelf (Crown Estate Commissioners 1983).

The second is the incorporation of new resources into the coastal and marine inventory as a function of technological advances (together, again, with the pressures on their land-based equivalents or alternatives). The offshore exploitation of oil and gas illustrates one stage in this process. Another is illustrated by the discovery of marine mineral deposits such as manganese nodules. These remain, for economic reasons, outside the present definition of resources but provide, nevertheless, potent incentives for addressing future systems of allocation and control.

The third is the discovery or recognition of new resources *per se*, where, with changing conditions, objects previously perceived as valueless, dispensable, or perhaps even unknown, are elevated to the realm of a resource. Additional marine species are, for example, considered as sources of protein, mudflats are recognized as important wildfowl areas or marine breeding grounds, and coastal waters are identified as potential sites for offshore islands or focuses of new forms of coastal recreation such as wind or para-sailing.

This increasing intensity of demand is allied with increasing evidence of conflict between the different uses and users of the coast and seas. It is similarly allied with an increasing awareness, and indeed numerous illustrations, of the finite, frequently fragile, always interconnected nature of the coastal and marine eco-systems, together with the associated, inherent limits to their carrying capacity. There is also a growing awareness of the potential for further conflict in the future if multiple use continues to expand in an unregulated manner.

We are witnessing, therefore, an essentially demand-related rise in the profile of the two areas. The coastal zone, an area encompassing land, shore and nearshore waters, is increasingly recognized as a functional region (see UN 1982). It is increasingly a feature of common parlance, a function of its growing importance to mankind.

We are also witnessing the corollary, the requirement for, and the methods of, apportioning and regulating these scarce resources. Because the coast and the seas have traditionally lacked the importance of land they also tend to lack their own well developed managerial framework. For the same reason they have also played only a minimal role in influencing the administrations which have developed to discharge managerial and planning functions on land. The rise in importance of coastal and marine resources therefore finds the administrative framework often either absent or peripheral to this new and revised focus of concern.

In the political power stakes the sparsely colonized area, rich in resources, does not, however, remain virgin territory for long. Agencies already active at the margin seek to extend their remit. Agencies with existing powers lying dormant, or undeveloped, seek to consolidate their influence and adopt a more vigorous approach. New agencies are formed to champion particular claims or focus attention on particular areas. Others, with well represented and vested interests, seek to maintain the *status quo*.

The need to examine our traditional approaches to management, to address the mechanism for dividing and discharging responsibility and for determining strategy, and indeed to assess their relevance to contemporary need, is therefore nowhere more evident than in this rapidly evolving realm.

CRITICAL ASPECTS OF THE CURRENT SYSTEM OF MANAGEMENT

Managerial systems can be divided essentially between those which are defined by area, that is, those which focus on a complex of resources in a particular locality, and those which are defined by the resources themselves. The dominant managerial framework in the coastal and oceanic environments remains resource or use-specific.

Different agencies are typically charged with responsibility for navigation, for the regulation of fisheries and for the conservation of wildlife; with responsibility for allocating the right to extract oil and gas or to dredge for minerals, to dump waste or to prevent pollution, even to protect coasts from erosion as opposed to protecting lands from flooding.

The basic framework is then compounded by varying spatial, hierarchical and functional divisions. In the English and Welsh example of fisheries management for instance, interests at the national level fall primarily within the ambit of the Ministry of Agriculture, Fisheries and Food (MAFF), which is responsible for fishery policy, research and advice. The Department of Transport is also involved, being concerned with the safety aspects of the industry, whilst the Ministry of Defence, through the Royal Navy and Royal Air Force, provide fishery protection, and the Foreign and Commonwealth Office helps with international negotiations.

Local interests are upheld by sea fisheries committees, which divide the coastline into twelve districts. These range in extent from one county, as is the case, for instance, in Devon, to five counties in the case of the Lancashire and Western Sea Fisheries District, which extends from Cardigan to Morecombe Bay. The regional water authorities also have a role to play and a different spatial basis, controlling salmon and trout. With the eventual introduction of Part II of the Control of Pollution Act 1974 they will also take over the role of the sea fisheries committees in controlling discharges to coastal waters.

The depletion of stocks and the tendency towards increasing exclusivity have also led to an increased international dimension as countries have sought to redistribute, or maintain, traditional interests and negotiate a new order. In respect of England and Wales the EEC now, therefore, plays an important role in allocating fishery resources and determining resource quality, as has the North East Atlantic Fisheries Commission in the past. The jurisdiction of fishery limits in relation to such administrative arrangements is regionally very complex.

In instances where a resource is common to both land and sea this divide is also frequently reflected in the discharge of responsibilities. Again, to use examples from England and Wales, the extraction of aggregates down to the low water mark, that is, including the foreshore, is controlled by planning permission and related conditions issued by county councils, although the Crown Estate Commissioners actually hold title to the foreshore. The commissioners also claim the territorial sea bed beyond low water mark, and in this latter area it is they who license the extraction of aggregates.

This dual loci of control can pose problems for long-term planning and resource management, particularly as the commissioners have traditionally had no policy for initiating exploitation but have rather welcomed proposals. This lack of strategy, and their financial involvement, contrasts with the procedures laid down by the planning system on land.

Responsibility for oil pollution is similarly divided. Here, subsequent to the *Torrey Canyon* disaster, local authorities were requested, by circular, to clean up oil within one mile of the shore, or on shore. Responsibility at this level is then further divided between county and district councils, whilst port and harbour authorities are also involved. Offshore, responsibility was allocated first to the Department of Trade and then to the Department of Transport, which, through the Marine Pollution Control Unit, has also assumed overall responsibility for supervising the clean-up of oil spills along the coastline.

Management whose focus is a particular area, as opposed to a particular resource in an area, is less common. Potentially it encompasses a series of co-ordinated policies directed at specific localities and backed up by an effective means of implementation and enforcement. In practice the spatial framework tends to be established for some other purpose.

In the coastal zone, for example, local government areas provide a potentially comprehensive, although often dormant, example. On land they are an important planning and managerial agency, able to exert, moreover, a pervasive influence over the activities of other organizations. Their jurisdiction tends to stop at low water mark but frequently possesses an additional maritime dimension, relating, for instance, to port or harbour functions, recreation, fisheries, aggregates and hydrocarbons (see, e.g. Halliday 1987). On examination this maritime dimension emerges, however, instead as a series of separate dimensions. Extensions have been drawn in response to one requirement alone, the remaining powers of local government have not been similarly extended, and the result is again therefore a series of offshore boundaries related to the control of individual resources.

A second type of areal management focuses on key areas where particular processes have combined to raise the profile of the area beyond a critical threshold and engendered a local response. Designation is frequently consequent on crisis.

The threats to the internationally important wetland, the Wadden Sea, for instance, have led to combined studies by the countries bordering the sea, Denmark, Germany and the Netherlands. One result has been the resolution that the three countries need a common policy and close co-operation in management, including the zoning of all human activities, if the integrity of this coastal resource is to be maintained. Another was the assertion that information on all aspects of the area was already available, yet individual authorities charged with its management were either insufficiently informed or not 'assisted' enough in the decision-making process (Wolff 1977).

Integrated and effective management relies, therefore, as the next sections emphasize, not only on the evolution of appropriate managerial structures and processes but also on changes in the human environment. Managers need increasingly to be aware of the interrelated and inter-sectoral nature of their operating environment and of the limitations to their traditional professional domains. The political will to consult, compromise and make full use of the procedures for joint action which already exist, is imperative.

The system is not, however, static. The increased recognition of coastal and maritime resources has, in effect, as suggested above, opened up new arenas where agency relations are re-examined and redefined.

Three particular developments need stressing. The first is the series of United Nations conferences on the law of the sea, particularly the most recent, UNCLOS III. Again well documented, this aimed to produce a treaty dealing with ocean space as a whole and, irrespective of ratification, represents probably the most significant single contribution to the emerging body of law of the sea. Aspects covered include naviga-tion, the exploitation of living and non-living resources, conservation, waste disposal, military uses and scientific research.

The assignation of responsibility to individual organizations is not, however, addressed, and executive responsibility remains with existing international organizations and individual nations. Whilst progress has therefore been made in explicitly addressing the issues and defining a basic legislative framework, the essential question of how these issues are to be integrated remains largely ignored.

Meanwhile the focus of this renewed phase of boundary demarcation is largely coincident with the oceans beyond the territorial seas. There remain therefore, the still maritime environment and resources of the territorial seas and internal waters, together with the coastal resources, all largely bypassed by this movement towards more holistic manage-ment. Here individual nations remain the determinants of policy and, as already seen, integration is limited.

The second is a counter-force, reflecting this lack of international activity in the maritime margins and exemplified by the extension and adjunction of primarily land-based systems into coastal waters.

The powers of the Nature Conservancy Council were extended, for instance, by the Wildlife and Countryside Act 1981. This now enables 'any land covered continuously or intermittently by tidal waters or parts of the sea in or adjacent to Great Britain up to the seaward limits of territorial waters' to be designated a national nature reserve. Powers previously terminated at low water mark.

Another example is the various extensions of local government powers mentioned above. Many are historical but have an increased significance today not only because of increased pressures on the zone and hence a desire to act, but also because they represent the latent potential of these authorities to exercise a degree of control beyond their coastline once the need is perceived and the political will is present.

The third, again mentioned above, is the increasing number of area-specific initiatives. Their proliferation draws attention to the necessity for an appropriate spatial definition of coastal and marine areas, as opposed, or perhaps more correctly in addition to, the more commonly known legal framework.

Meanwhile, as this section has shown, coastal and marine responsibilities are discharged by an essentially tripartite system, whereby coastal land, nearshore waters and oceanic resources form distinct managerial provinces. This triumvirate is then superimposed upon a use-specific framework and compounded by a variety of spatial jurisdictions. The next two sections look at some alternatives to accepting the gradual evolution of this *laissez-faire* system, a system which owes much to the historical subservience of sea to land. In particular they propose a redefinition of the managerial system along the lines of functional regions, key uses and critical interactions.

DEFINITION OF COASTAL AND SEA REGIONS

The definition of coastal and sea regions is perhaps the most useful starting point for the integration of coastal and sea use management. The three principal sets of criteria to be considered are the legal, physical and social. Of these the legal criteria, expressed in terms of jurisdictional limits seaward, are the most straighforward, the principal seaward limits being those of internal waters, the territorial sea (up to twelve nautical miles), and the sets of EFZ and EEZ limits. To landward there are, of course, generally no limits of this kind, with the possible exception of byelaws governing the use of rivers in, for example, port areas. Land use planning systems and coastal inventories (on the French model) may extend to high or low water marks without clear distinction from landward management arrangements.

What is thus required is a clear system of use arrangements related to jurisdictional boundaries. The law of the sea has usually been relatively clear on this. The principal confusion lies rather with the seaward jurisdiction of land management organizations, such as local, port or river authorities. Beyond this, some idea of the detailed nature of the desirable transition is necessary. For example, it may be that, in general, internal waters are best considered part of the land for almost all purposes, except possibly navigation. The main challenge then lies in the efficient administration of the territorial sea, most probably set at twelve nautical miles, as in the UK case. Beyond this, management should probably conform to the Law of the Sea Convention of 1982 and its subsequent amendments.

The legal criteria must, of course, be related to the physical and social criteria upon which they are based. Regionally, seaward influences extend landward and vice versa. However, from a managerial point of view only the maritime jurisdictional zones are clearly defined.

The uses of the land and sea present a more difficult situation. The starting point is the measurement of intensity of uses. This is generally significantly greater to both the landward and seaward sides of the shoreline, with relatively high levels of both conflicts among uses and environmental impacts found in this relatively narrow zone. The territorial sea of twelve miles is probably, on the whole, suitable as a managerial unit, although there are some areas of intensive use beyond it, especially in the case of navigation, fisheries, mineral extraction and conservation, where special arrangements may be necessary. To landward it is often necessary to employ special planning restrictions on land use, especially to prevent the spread of urban land uses and maintain coastal conservation interests.

OUR TECHNICAL AND GENERAL MANAGEMENT APPROACH

Globally it is necessary to consider three major groups or combinations of land management arrangements. The first, the management of the land in urban areas, has been dominated for many years by the land planning system and secondarily by the valuation of space. Other management aspects, such as anti-pollution measures and water supply, have tended to be viewed as separate. In the second, the settled rural areas, full land use planning arrangements are liable to be absent and management restricted to certain environmental measures. The third major group of land environments, the extensive but sparsely settled or uninhabited wilderness regions, generally have little if any management system at all, except for conservation arrangements in some regions.

Sea use management arrangements have hitherto been akin to a cross between rural and wilderness land management categories. The most important are use-based organizations and regulations. These are both national and international in character, best developed in the traditional fields of navigation and fisheries, and exerting, until recently, the minimum of control beyond the territorial seas. Despite variations in sea use intensity, the marine environment is, as a whole, perhaps less use/environmentally differentiated than the three land environments, because of the properties of the water column, and a more uniform management system (based on jurisdictional zones) may therefore be possible.

Because of this lack of environmental differentiation and the relatively undeveloped state of sea use, it is probably also easier than on land to consider, and even develop, an integrated management system based on the interactions of man and the sea (Smith 1985). The starting point for such a system is an understanding of the precise extent and nature of man/sea interactions, which, in turn, are related to the fundamental purposes for which man uses the sea. These purposes can be summarized in eight use groups. These are illustrated in Figure 12.1, together with their key interactions.

The interactions apply, in various ways, to all the use groupings. Monitoring, surveillance and information technology, for instance, are essentially the scientific input into management, as well as being a major objective. Technology assessment and project development are the technological input of the engineers. The environmental assessment grouping includes risk assessment applied both to the physical and economic environments, the valuation of resources and the relatively new field of environmental impact assessment.

It is possible to categorize land use groupings and interactions in a similar manner. To the eight fundamental use groups should probably be added coastal engineering (the coast is one of the most modified natural environments in many instances); human settlement and industry. The management-related interaction categories remain constant, although for the most part, they are much better developed on land than at sea. From a technical management standpoint, the challenge of effective coastal management is then first to recognize the essential symmetry in man/environment relationships which are strongly reflected in both technical operations and management organization, and second to make special provision for the extra land use (i.e. social) categories in the coastal zone.

Beyond technical management lies the general management function. The two fundamental operations in a maritime context are arguably the co-ordination of technical management functions, on the

Technical management fields

		Science		Technology		Environmental assessment			Social assessment		

1 Information technology
2 Monitoring
3 Surveillance
4 Technical assessment
5 Project development
6 Hazard assessment
7 Resource assessment and valuation
8 Environmental impact analysis
9 Sea/land use planning
10 Law: sea and coastal zone
11 Social impact assessment

General management fields

12 Co-ordination of technical management
13 Policy/planning

Sea use fields
1 Navigation and communication
2 Strategic
3 Mineral and energy extraction
4 Fisheries and aquaculture
5 Waste disposal and pollution control
6 (Marine) recreation
7 (Marine) environmental science
8 Conservation

Additional land use fields
9 Coastal engineering
10 Human settlement
11 Industry

Figure 12.1   A management matrix for sea and land

one hand, and strategic policy formulation on the other. The key to effective coastal management probably lies in identifying the 'correct' sets of land/sea environments with distinctive combinations of both general and technical management characteristics. These environments may be made up from the aforementioned complexes of the territorial sea and internal waters, together with the three sets of land environments outlined above: urban areas, rural settled regions and the sparsely inhabited or uninhabited wilderness.

With such perspectives it is then possible to distinguish the degrees of management development of both land and sea. In fact only urban areas, rural areas and internal waters/territorial seas are managed to any extent at all, especially in the sense of land use planning, that is, the spatial allocation of resources. Most of the technical management categories also fall into the regions. Only a few, especially monitoring and surveillance, and hazard/resource assessment are practised to any extent beyond, in the land wildernesses and seaward of the territorial sea (EFZ/EEZ).

One ensuing challenge is to formulate effective management strategies for the longest coastal stretches of all, that is, the world's undeveloped coastal regions. In the more intensively used areas the challenge is to apply common technical and general management principles *across* the shoreline, taking into account the distinctive coastal combinations. For example, it may be that the southern North Sea/eastern English Channel is one such area.

PRACTICAL CONSTRAINTS IN THE PRESENT SYSTEM

Various features of the present situation still militate against the adoption of such an idealized scheme. They can be divided into structural limitations and human or perceptual limitations.

The first of the structural limits is the form of the prevailing administrative systems. This obviously varies from one country to another but essentially the contemporary divided organization militates against an awareness of the true scale and interrelated nature of coastal and oceanic problems. In turn only an appreciation of the scale and reciprocity of these problems can cause a review of administration and priority.

One potential solution is the further formation of multi-organizations. These are composites drawing on several organizations and possessing functions ranging from a forum for the exchange of information to a body with considerable delegated powers. This enables wide-ranging interests to be drawn together with a view to co-ordinated management, whilst also allowing the area to remain within the purview of individual participants. The extra dimension within such areas can therefore merge with the existing organizational structure and so avoid many of the problems associated with the creation of a new agency.

This is indeed already happening in many of the key areas mentioned above, where multi-organizations have been formed in an attempt to provide integrated management. The Solent Sailing Conference, the Langstone Harbour Board and the Chichester Harbour Conservancy are all good examples from England's congested south coast. Limitations, as with any theory, are, however, apparent and several factors militate against their effectiveness in practice. The latter include the ability of the participants to appreciate the common task and forsake their own agency priorities and the ability of even a forum to represent the views of all interested parties.

A second structural limitation relates to the availability of knowledge. Increased information is particularly required on the relationship between coastal processes and use, to allow the determination of tolerance thresholds, and also on the relative importance of individual resources and activities, in order that their coastal or marine claim can be better assessed. More informed decision-making should make better decision-making.

Much information is, of course, already in existence but, as already noted, inadequate dissemination means resources are wasted in duplicating the learning process, often to the extent of creating environmental or economic disasters which could have been avoided if existing knowledge had been made available or taken into account. Again, avenues for disseminating the information need to be developed, as is clear from our managerial matrix, but we also return to the persistent theme of institutional and perceptual barriers to communication and hence to informed action.

Participating authorities, because of the desire to maintain or reinforce autonomy, to preserve traditional areas of singular expertise, or to consolidate the integrity or value of a profession, are reluctant to cede control and this includes making information available. Information is a strategic weapon in the power stakes of the coastal and oceanic environment. It could be a far more effective weapon if resources were pooled. Attitudes therefore lie again at the bottom of what is also a structural problem.

A third structural problem is the need for a final source of arbitration. There remain very real conflicts in deciding resource priorities. The problem is more than one of different strategies which require ordering; it is also one of different viewpoints and value systems.

Consensus can never be obtained. Decisions need, therefore, to be taken which lie beyond the art of compromise, and this is probably where the major requirement for structural change exists.

On the human side the absence of political will remains a major impediment to change. Only if concern for the coast and the oceans is set on a higher plane can the chain of positive feedback be set in progress, to ensure that the area receives increased priority and enhanced management. Public opinion remains the essential underlying force; it fuels such action and is the body to which such action is ultimately accountable. Education is therefore fundamental.

Education can also change professional attitudes. It needs, for instance, to reduce antagonism towards the administrator who draws on several specialisms and to imbue sectors with an awareness that coastal problems need to be solved by a merging of minds. The process of consultation must be seen not to threaten the interests of the contributors but rather to increase the effectiveness with which they can discharge their own responsibilities. Joint action must be seen as politically attractive.

CONCLUSION

The first point to make is that, despite a relative lack of coastal zone management arrangements *per se*, there is, in the developed world at least, a large and complex system of coastal zone management in existence. For many purposes this is going to be the starting point for a large part of the management of the sea. It is supplemented, for the marine environment, by the existing and considerable framework of use-based sea use management arrangements. The immediate need is to work from these existing arrangements towards a common, technical management-based approach, as suggested in our model, which will initially cover internal waters and the territorial sea.

This approach requires clarification of administrative jurisdictional boundaries, especially those of land-based organizations. Beyond this, it is probably easier to begin a fully integrated management approach with the sea rather than with the land, working backwards, as it were, towards the coast. Here the notion of general management is crucial, clearly distinguishing much of the routine management work from the strategic thinking and organization behind it, and yet providing a common set of principles applicable to both land and sea. Such an approach will have practical working and economic implications, permitting the efficient use of management resources, be they scientific, economic or human.

It also requires an increased willingness to actually recognize and address the problems that face our coastal and marine areas. At least we are now beginning to make progress in elevating the subject on the political agenda.

REFERENCES

Clark, R.B. (1987) *The Waters around the British Isles: their Conflicting Uses*, Oxford: Clarendon Press.

Couper, A.D. (ed.) (1983) *The Times Atlas of the Oceans*, London: Times Books.

Crown Estate Commissioners (1983) *The Crown Estate (1983): Report of the Commissioners for the Year ended 31 March 1983*, London: HMSO.

Great Britain, Parliament Public General Acts (1981) *Wildlife and Countryside Act 1981*, chapter 69, London: HMSO.

Halliday, J.E. (1987) 'Beyond the bounds? A consideration of local government limits in the coastal zone of England and Wales'. in G. Blake (ed.) *Maritime Boundaries and Ocean Resources*, London: Croom Helm.

Smith, H.D. (1985) 'The management and administration of the sea', *Area* 17 (2): 109–15.

United Nations Ocean Economics and Technology Branch (1982) *Coastal Area Management and Development*, Oxford: Pergamon Press.

Wolff, W.J. (ed.) (1977) *Proceedings of the Conference of Wadden See Experts*, Copenhagen.

13 Sea use planning and fish farms in Scotland

A case study *

E. Earll, A. Ross and S. Gubbay

INTRODUCTION

As the demand for use of the coastal zone has increased over the years maritime nations have developed a variety of strategies for coping with the situation. In most cases these are heavily reliant on a system of planning which aims to balance the use in a way which satisfies the requirements and responsibilities of the country. However, whilst trying to cope with these aspects, it is clear that the pressures have extended beyond the coastal strip and the foreshore, and now include both inshore and offshore waters. In the 1980s, with the proliferation of exclusive economic zones, nations also have more jurisdiction over areas of sea. The need for sea use planning is therefore becoming more widely accepted.

In the UK the management of the coastal strip, the inshore waters and territorial sea is a complex issue. The planning provisions are very variable, depending on the activity under consideration. For example, with regard to coastal oil and gas-related developments the Scottish Development Department has drawn up planning guidelines identifying Preferred Conservation Zones and Preferred Development Zones (Scottish Development Department 1974). In contrast, there are few clear guidelines for the development of fish farming. The opportunities for planning are further confused by the many pieces of legislation which are relevant to the coastal zone and the number of government departments involved (Tables 13.1–2) and it has often been suggested that there is a case for a single Department of Maritime Affairs with responsibility for marine resources (Shaw 1983).

The environmental problems resulting from particular developments are another aspect which need to be considered in the planning

* There have been many important developments since this chapter was submitted in 1988.

Table 13.1   Government departments with responsibilities in the UK coastal zone

Crown Estate Commissioners
Department of Agriculture and Fisheries for Scotland
Department of Education and Science
Department of Energy
Department of the Environment
Department of Trade and Industry
Department of Transport
Foreign and Commonwealth Office
Health and Safety Executive
Home Office
Ministry of Agriculture, Fisheries and Food
Ministry of Defence
Natural Environment Research Council
Nature Conservancy Council
Science and Engineering Research Council
Scottish Development Department
Welsh Office

framework. An EEC directive (EEC/85/337) specifies projects where an environmental impact assessment will be mandatory prior to approval (Annex I projects) and a second list of projects where assessment will need to be carried out only 'where Member States consider that their characteristics so require' (Annex II projects). Projects that are likely to take place in the coastal zone include oil refineries and nuclear power stations (both Annex I projects) and marinas (Annex II).

These problems of sea use planning, the range of organizations responsible for activities in this area, and the need to take the environmental impact of developments into account can be demonstrated by using the example of the development and current status of marine fish farming in Scotland.

BACKGROUND TO THE FISH FARMING INDUSTRY

Fish farming is a rapidly expanding industry all over the world, following technological advances in production techniques in the 1970s. Fish farming in Scotland has grown enormously in recent years, particularly the cage farming of the Atlantic salmon, *Salmo salar*. The other main elements of the Scottish fish farming industry are the production of the rainbow trout *Salmo gairdneri*, mussel, *Mytilus edulis*, Pacific oyster, *Crassostrea gigas*, and queen scallop, *Chlamys opercularis*.

Salmon production is concentrated mainly in the confined and

Table 13.2 Legislation relevant to the UK coastal zone

Carriage of Goods by Sea Act 1971
Civic Government (Scotland) Act 1982
Coast Protection Act 1949
Conservation of Seals Act 1970
Continental Shelf Act 1964
Control of Pollution Act 1974
Criminal Justice Act 1948
Criminal Justice Act 1982
Crown Estate Act 1961
Crown Lands Act 1866
Crown Lands Act 1906
Customs and Excise Act 1952
Customs and Excise Duties (General Reliefs) Act 1979
Customs and Excise Management Act 1979
Dangerous Vessels Act 1985
Deep Sea Mining (Temporary Provisions) Act 1981
Diseases of Fish Act 1937
Diseases of Fish Act 1983
Dumping at Sea Act 1974
Explosive Substances Act 1875
Fisheries Act 1981
Flood Prevention (Scotland) Act 1961
Food and Environment Protection Act 1985
Harbour Transfer Act 1862
Harbours Act 1964
Health and Safety at Work Act 1974
Highways Act 1959
Hovercraft Act 1968
Illegal Trawling (Scotland) Act 1934
Inshore Fishing (Scotland) Act 1984
Interpretation Act 1978
Land Drainage Act 1976
Land Drainage (Scotland) Act 1958
Land Powers (Defence) Act 1958
Land Registration Act 1925
Land Registration (Scotland) Act 1979
Local Government Act 1972
Local Government (Miscellaneous Provisions) Act 1976
Local Government, Planning and Land Act 1980
Local Government (Scotland) Act 1973
Malicious Damage Act 1861
Merchant Shipping Act 1894
Merchant Shipping Act 1970
Merchant Shipping Act 1974
Merchant Shipping Act 1979
Merchant Shipping Act 1983
Merchant Shipping Act 1984
Merchant Shipping (Oil Pollution) Act 1971

Table 13.2   Legislation relevant to the UK coastal zone

Military Lands Act 1900
Mineral Working (Offshore Installations) Act 1971
Nature Conservancy Council Act 1973
Offshore Petroleum Development (Scotland) Act 1975
Oil and Gas Enterprise Act 1982
Oil and Pipelines Act 1985
Petroleum Act 1987
Petroleum (Production) Act 1935
Petroleum and Submarine Pipelines Act 1975
Pilotage Act 1913
Pilotage Act 1987
Pipelines Act 1962
Prevention of Oil Pollution Act 1971
Protection of Wrecks Act 1973
Public Health Acts Amendments Act 1907
Public Health Act 1936
Public Health Act 1961
Radioactive Substances Act 1960
Rivers (Prevention of Pollution) Act 1961
Rivers (Prevention of Pollution) (Scotland) Act 1951
Road Traffic Regulation Act 1984
Salmon and Freshwater Fisheries Act 1975
Salmon and Freshwater Fisheries (Prot.) (Scotland) Act 1951
Science and Technology Act 1965
Scottish Development Agency Act 1975
Sea Fish (Conservation) Act 1967
Sea Fisheries Act 1968
Sea Fisheries (Shellfish) Act 1967
Sea Fisheries Regulations Act 1966
Seal Fishery Act 1875
Seal Fisheries (North Pacific) Act 1895
Seal Fisheries (North Pacific) Act 1912
Spray Irrigation (Scotland) Act 1964
Territorial Sea Act 1987
Water Act 1983
Whaling Industry (Regulation) Act 1934
Wildlife and Countryside Act 1981

---

sheltered sea lochs of the north and west coast of Scotland. The two main island groupings (the Western and Northern Isles) each produce about 15 per cent of the total Scottish production, with the remaining 70 per cent being grown on the west coast of the mainland and the inner isles.

The initial development of the Scottish salmon farming industry has taken place as a result of substantial investment in research and development since the late 1960s by multinational companies, notably Unilever, Booker-McConnell, BP and Shell. Salmon production has increased enormously over the last decade from 520 tonnes in 1979 to 2540 tonnes in 1984 and 12 700 tonnes in 1987 (Department of Agriculture and Fisheries for Scotland (DAFS) figures), having a first sale value of around £50 million. Predictions from DAFS suggest that production would reach 31 000 tonnes by 1989 and rise to 52 000 tonnes by 1991.

Future developments are sure to include growth in the cultivation of other marine species, such as turbot, cod and halibut, which are the subject of successful early trials and on-going research. Considerable growth is expected also in the shellfish sector, where much research is under way. For example, UK production of mussels is expected to rise from 3350 tonnes (1986) to 14 000 tonnes in 1991 (Ministry of Agriculture, Fisheries and Food (MAFF)/DAFS).

This growth has been facilitated by substantial government support. The Highlands and Islands Development Board (HIDB) provided grants to assist the development of fish farming, totalling £20 million by 1987 (concentrating especially on small operations since mid-1985), with approximately £44 million of private-sector investment. HIDB justifies its grant-aid expenditure chiefly in terms of employment created or protected in remote areas with otherwise poor employment opportunities. In 1987 there were 126 companies producing Atlantic salmon at 207 sites, employing 608 full-time and 198 part-time staff (DAFS, unpublished figures). In addition, many more 'downstream' jobs are provided in fish processing, boat-building, marine engineering, net and cage manufacture, feed compounding and haulage.

ENVIRONMENTAL PROBLEMS OF FISH FARMING

Marine fish farming is a use of natural resources that clearly benefits the local economy, especially in remote areas. However, the rapid growth of the industry raises real concern as to its adverse effects, most of which are not, as yet, quantified or fully understood.

Pollution

This takes a number of forms:

1 Organic pollution of the benthos by faecal and waste food material.
2 Nitrogenous waste, causing nutrient enrichment.
3 Chemical inputs from farm management such as anti-foulants on nets and cages, chemotherapy for the treatment of fish diseases and parasites, disinfectants, and food additives such as antibiotics and colour enhancers.

4 Waste disposal and litter in the form of dead and diseased fish, disused nets, polythene feed bags and broken-down polystyrene floats.

Sea lochs are being increasingly developed without adequate knowledge of their carrying capacity, in terms of the immediate or cumulative effects of these inputs on the eco-system.

Effects on wild salmon

Genetic degradation

The escape and release of farmed fish pose a threat to wild salmon populations by introducing artificially selected features such as more placid behaviour, later sexual development and faster growth, which can depress the performance of the wild fish.

Disease and parasites

The high incidence of disease and parasite infestation in farmed fish, against which they are regularly treated with drugs and chemicals, will increase the infection rates in wild salmon which are not protected.

Predator control

Wild birds and mammals which are natural predators of fish and shellfish are attracted to fish farms as a source of food. These are principally grey and common seals, herons, cormorants, shags, otters and mink on salmon farms and eider ducks on mussel farms. Although fish farm protection usually includes the use of nets and often deterrent devices, predators are commonly killed by shooting, trapping, entanglement (drowning) or poisoning.

Impact on sea fish stocks

The fish farming industry is making increasing demands on the 'industrial' stocks of fish, such as sandeel, which are used in the manufacture of fish feed pellets.

In addition to ecological and wildlife effects, the fish farming industry impinges on a number of other aspects of the environment and its users. It is these effects that have attracted most attention (e.g. Cobham Resource Consultants' report to the Countryside Commission for

Scotland). This illustrates a common problem with marine developments, where the emphasis tends to be on the visual impact.

Landscape, tourism and recreation

Fish farms are typically sited in areas of very high scenic value, and there is concern that the industry could be detrimental to the economy of some areas by reducing the 'wilderness' quality and thus the potential for tourism.

Navigation and other water use

Fish farms are frequently sited in narrow or confined bodies of water and may thus present obstructions to access or safe anchorage.

Social considerations

Current policy for the allocation of sea bed leases has allowed farmed salmon production to be dominated by a few large multinational companies. While their valuable contribution to research and development in the industry is acknowledged, the situation is likely to have adverse implications in ecological and social terms. It limits the opportunities for the potential small local operator, investment is transferable internationally according to existing and potential profit levels, high technology and economies of scale actually reduce the employment opportunities, and the majority of profits are usually exported from the area.

The range of problems, together with the incentive to develop the new industry, suggest the need for a co-ordinated, planned approach. However, the fish farming industry demonstrates considerable shortcomings in this respect. These are highlighted in the following sections, which deal with the planning, regulation, legislation and organizational problems of fish farming.

PLANNING AND REGULATIONS OF FISH FARMING

Development stages

As with the development of any major new industry the fish farming industry has moved through a number of stages.

Initial stages

The fish farming industry started to take off on a commercial scale about ten years ago, and the Scottish Office clearly felt that this activity offered enormous potential for the coastal areas of western Scotland, which have been in decline for many years. To support this DAFS played an important part in negotiations for European funds from the Integrated Development Programme (IDO) to be channelled into the industry through the HIDB.

At this stage there was very little regulation by any of the agencies which have duties in this sphere. For example, until October 1986 fish farming discharges were exempt from the requirement for River Purification Board (RPB) consent; after that date the exemption was modified, bringing tank farms and cage sites under RPB control. Similarly, until 1986 the Crown Estate Commissioners (CEC) charged a nominal £20 per site per year for sea bed leases for salmon farms. Subsequently the CEC was instructed to obtain an economic return from the leases and raised the fee to £50 per tonne per year.

Maturing

The industry in 1988 could be said to have passed though the 'honeymoon' stage of its development. By its sheer scale and extent the industry inevitably conflicts with a variety of other uses of the marine environment and they, in their turn, have exerted pressure on their respective statutory bodies or have complained publicly about the increasing impact of the industry.

This has prompted a surge of consultative documents and recommendations, particularly since 1987, which, in the long term, are likely to lead to a more closely regulated industry within a clearer operational framework.

Established industry

The industry has yet to reach the point where its maximum potential has been realized or where the planning and regulatory controls are satisfactory, or even in place, effectively to curb the worst environmental impact. As the industry becomes established new trends, especially towards the use of more exposed sites, will also pose new and different environmental problems.

National versus regional aspects

The government clearly supports the development of fish farming, as evidenced by the speeches of ministers and the development work of agencies such as the HIDB and CEC. However, there is no clearly defined policy statement available from which one can judge the government's intention. Government reports like the 'Multi-annual Guidance Programme for Aquaculture' (1987) indicate overall trends but again are inaccessible to the majority of interested parties. In addition, there is no clearly defined lead ministry for the industry and, evidently, no formal co-ordination of government departments at a national level.

At the regional level there is considerable frustration with the planning/consultation procedure which is currently being operated by the CEC. This is due to its lack of public accountability, an inadequate consultation procedure and the fact that it is administered from a central base. Whilst it may seem attractive to transfer duties for offshore planning to regional councils, most have no statutory remit beyond the seashore in Scotland. An exception is the Shetland Islands Council, which is empowered under the Zetland County Council Act 1974 to control all developments between the low water mark and the twelve mile limit. Other councils are pushing for more formal control. For example, Highland Regional Council has produced framework plans for fish farming in several sea loch areas. However, it would require either new primary legislation or modification of existing legislation for these to be implemented lawfully.

Demands are also being voiced for a more effective mechanism for assessing sea bed lease applications from shellfish farmers who want to develop bottom deposits for ranched scallops, etc.

Legacy from the initial stages

Regardless of how the planning and regulatory framework develops, unless retrospective legislation is introduced, which is extremely unlikely, there will be a number of long-term legacies of these initial stages. One of these is the fact that many sites are now leased to fish farmers for substantial periods of time (twenty to ninety-nine years).

LEGISLATION RELATING TO FISH FARMING

The duties of government departments are often clearly defined under particular Acts. However, in practice these duties may overlap with other departments', and liaison and co-ordination are seldom in evidence. A small number of examples will suffice.

The River Purification Boards have a clear duty to administer the Control of Pollution Act 1974 (COPA). Since the implementation of the modification of section 34(1) in October 1987, Shetland and Orkney Islands Councils and Highland RPB have seen fit to administer consents to fish farm operators, even though they break the environmental quality objectives set out in the RPB's policy guidelines. In contrast, the Clyde RPB questioned whether COPA could in fact be applied to offshore fish farms, considering that its duty to issue consents applied only to shore-based discharges. After taking legal advice the Clyde RPB began issuing discharge consents to fish farms in 1988.

Another example of confusion relates to the control of chemicals used by fish farms, e.g. Tributyltin (TBT) or Nuvan. These have a very ambivalent position within the statutory control system. TBT anti-foulant was initially controlled under COPA but is now regarded as a pesticide and comes under the Food and Environment Protection Act 1985 (FEPA); this falls under the remit of DAFS. Shellfishermen in particular have found the position of DAFS over TBT and salmon farming very difficult – indeed, at least two cases of litigation are now under way because of destruction of shellfisheries caused by TBT on salmon farm nets.

Nuvan is an extremely toxic treatment for sea lice, which present a major problem to the industry. There are no statutory guidelines for its use in this application and it has been prescribed by vets under the Medicines Act 1968 although it has not yet been licensed for this use. There have been a number of incidents where poorly controlled applications of Nuvan have resulted in large numbers of the salmon being killed.

In both Norway and Canada the 'carrying capacity' of bodies of water has been a major factor in the planning and regulation of the fish farming industry. This is only now being addressed by research in UK (CEC press release, February 1988) and seems extremely unlikely to figure in any practical way in the planning or regulation process in Scotland in the near future.

Similar grey areas exist with regard to a wide range of environmental issues, and this fact, coupled with departmental intransigence, does not bode well. For example, in relation to predator control, Ross (1988) has shown that significant numbers of seals and other predators are being destroyed each year by the industry. Predation represents a financial loss to the industry of between 2–4 per cent of total turnover. DAFS is the government department with a duty to provide advice and research on predator control but it has no written advice on this for the industry. It has no up-to-date information on the scale of the problem nor any plans to collect any. The Marine Conservation Society approaches this question with a wildlife bias, but neither the environment nor the industry is well served by this level of activity or commitment.

ORGANIZATIONAL PROBLEMS

As with many sea use planning issues there are a large number of government departments involved with the fish farming industry; they are listed in Table 13.3.

Table 13.3 Environmental issues and the related government departments and legislation

Issue	Department	Legislation
Chemical pollution		
Sea bed organic pollution	RPBs	COPA 1974
Nutrients	RPBs	COPA 1974
'Chemicals'	RPBs	COPA 1974
Pesticides	DAFS	FEPA 1985
Veterinary chemicals e.g. Nuvan	DAFS/MAFF	Medicines Act 1968
Use of chemicals at work	HSE	HSE 1974
Reduced oxygen	RPBs	COPA 1974
Species		
Diseases	DAFS	Diseases of Fish Act (1983)
Non-commercial species	NCC	Wildlife and Countryside Act 1981
Alien (introduced) species	DAFS/MAFF	W&C Act 1981
Shellfish	DAFS/MAFF	Various fisheries Acts
Commercial fish	DAFS/MAFF	Various fisheries Acts
Salmon	DAFS	Salmon and Freshwater Fisheries Act 1975
Seals	DAFS/NERC	Conservation of Seals Act 1970
'Pests' (mammals/birds)	DAFS	W&C Act 1981
Areas of sea		
Marine nature reserves	NCC	W&C ACT 1981
Special protection areas	NCC	EC bird directive
Sea loch 'carrying capacity' *re* fish farming	DAFS	None
Marine consultation areas	NCC	None
Very sensitive areas	CEC	None

The larger departments

As is often the case with marine issues, whilst departments often administer specific legislation there are many areas of overlap of both 'duty' and 'interest', where a particular issue is of common interest to one or more departments. For example, TBT paint was of concern to the RPBs as a 'noxious chemical' under COPA, of concern to DAFS and shellfishermen because of its effects on oysters, and now currently FEPA, and to the NCC because of its damaging effects on a range of marine wildlife. In fact, even though it was known to be a highly toxic substance, it was not until mid-1987 that anything was done to control its use.

This example is typical of the environmental issues which fish farming has raised and it illustrates how, even with primary legislation in force, there is both confusion of responsibility and a market reluctance to counter a well described and researched environmental threat.

Smaller agencies

There are a wide range of smaller government agencies with interests in fish farming, such as the Natural Environmental Research Council (NERC), with duties under the Conservation of Seals Act 1970, and the Sea Fish Industry Authority, which promotes the development of the industry and is also heavily involved in research. HM Coastguard administers the safety of inshore coastal traffic and has found that unchartered, unlit fish farms can be a major obstruction to coastal vessels.

Other organizations

There are a number of other organizations, covering a wide range of interests, which have been drawn into the fish farming debate. They include: environmental issues from pollution to predation, lack of planning, etc.; mobile gear fishermen, concerned about unlit, unmarked cages, obstruction, sea bed pollution; shellfishermen, concerned about CEC sea bed lease administration, chemicals, e.g. TBT, Nuvan; yachtsmen, concerned about occupation of safe anchorages, anchorages fouled with fish farm waste, unlit and unmarked cage sites; local residents, concerned about scenic disruption, noise, the killing of predators.

These organizations have not co-operated to any great extent in their attempts to make sense of the various government responsibilities, with the notable exception of the Scottish Wildlife and Countryside Link.

The Link initiative has succeeded in welding together the views of a wide range of organizations in an authoritative and widely available report (1988). The NGOs have often voiced their frustration at the lack of public accountability of agencies such as the CEC, or the apparent bias of government agencies against their activity. Arguably, until these groups can co-operate more effectively to exert sustained pressure, little will be done by government to alter the current situation.

ENVIRONMENTAL IMPACT ASSESSMENTS AND FISH FARMING

Environmental impact assessments (EIAs) are a useful tool for sea use planning, and procedures are rapidly developing, at both an administrative and a scientific level, for EIAs under the impetus of the EC directive mentioned above. A body of extremely valuable documentation already exists in this regard (HMSO 1981) and outlines procedures which could be followed in relation to fish farming (Figure 13.1).

Currently the CEC's procedures for the assessment of sea bed lease applications go little further than the first box ('Acquisition of information'). Although since October 1986 there has been an extensive system of consultation involving over twenty-five interest groups, there is no evidence of a systematic evaluation of comments or objections. There is no documentation of CEC's policy or procedures and there is no public presentation of the results, nor any system of appeal.

The current consultation procedure does not allow enough time for the consultees to do more than a desk study of particular sites. There is no overall planning strategy, and the consideration of sites is undertaken on an individual basis, with little regard for the carrying capacity of the sea loch as a whole. There seems to be little or no consideration of the widespread effects of pollutants, which may extend well beyond the limited area covered by the sea bed lease. Indeed, the CEC seems have great difficulty in understanding that there can be damaging environmental effects beyond the cage site.

There is no onus on the developer to undertake any EIA work, except in the case of extremely large developments. Indeed, the smaller operations would find such work difficult to fund, although this is not true of the large 'multinational' operators which account for sixty–seventy per cent of the production of the industry in 1987.

The EC EIA directive does not seem to offer any solution to the multiplicity of small developments of similar type which individually may represent little or no threat but which, in cumulative and national terms, represent a major environmental problem.

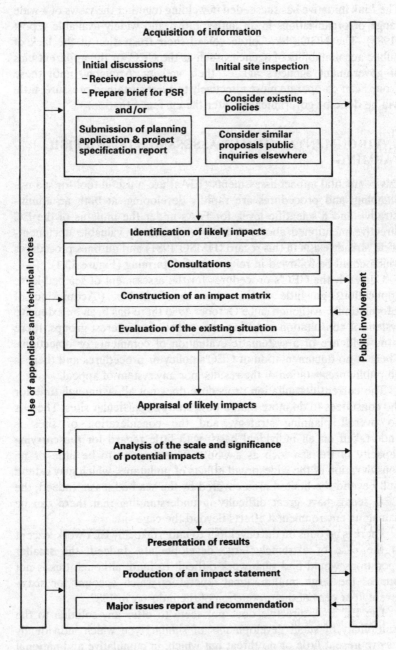

Figure 13.1 Linked activities in the appraisal method (from HMSO 1981).
Crown copyright reserved

Once again the need for a strategic approach at both a regional and a national level becomes apparent, and yet the current responsible agencies, in their present form, possess neither the resources nor the operational style (co-operation and openness) which are required for this process to work in practice.

CONCLUSIONS

This chapter has described, using the current developments of the fish farming industry in Scotland, some of the problems which are arising in relation to sea use planning in the UK to cope with the increasing pressures on the coastal environment. It is quite clear that any developing industry in its earliest stages is unlikely to welcome excessive regulation. However, given the substantial government backing of the industry, it is surprising that a wide range of environmental issues should still be unresolved at this stage in its development. Whilst there is strong commitment to the development of fish farming by many government departments, the lack of a clearly defined national strategy is counterproductive.

It is extremely unlikely that there will ever be a government department of marine affairs, which would, ideally, have supervised the development of fish farming. However, a device which was used with considerable success for the 1987 North Sea ministerial conference was the nomination of a lead minister and department to co-ordinate the activities of different government departments. This device could certainly be introduced productively to meet the demands of the developing fish farming industry.

What is advocated is the introduction of forward planning and development control by local authorities, based on a full assessment of environmental impacts and backed up by national planning guidelines.

REFERENCES

HMSO (1981) *A Manual for the Assessment of Major Development Proposals*, London: HMSO.

Ross, A. (1988) *Controlling Nature's Predators on Fish Farms*, Ross on Wye: Marine Conservation Society.

Scottish Development Department (1974) *North Sea Oil and Gas: Coastal Planning Guidelines*, Edinburgh: Scottish Office.

Scottish Wildlife and Countryside Link (1988) 'Marine Fishfarming in Scotland', discussion paper by the Fish Farm Working Group.

Shaw, D.F. (1983) *Conservation and Development of Marine and Coastal Resources*, Part 4, *World Conservation Strategy*, London: Kogan Page, pp. 261–314.

14 Integrated Fisheries Management models

Understanding the limits to marine resource exploitation

J.M. McGlade and R. McGarvey

This chapter presents the conceptual and theoretical basis of Integrated Fisheries Management (IFM) models, and uses as an example the haddock fishery off the south-west coast of the Atlantic Canadian province of Nova Scotia. Hypothesizing a predator–prey cause for the fifteen year cycle in catch that began in 1960, we construct a model that includes both biology and economics. In a series of Monte Carlo simulations the effects of constant and stochastically varying recruitment are examined. Further, we explore management strategies that aim to damp the cycle of fish-stock abundance and hence provide a more stable economic base for fishing communities. Fixed limits on effort and cost, as well as time-dependent management strategies which respond to fish-stock size and profit, are investigated. The results in this simulation system reveal that a strategy of fixed upper limits on effort identical to the classic equilibrium prescription of Gordon stabilizes the fifteen-year cycle. IFM models incorporating both ecological and socio-economic aspects of a fishery can thus be used to define strategies in which the expectations and goals of fishermen and managers converge.

INTRODUCTION

Integrated Fisheries Management and the global regime

Fisheries management is prefaced by the need to restrict access to resources. But in so doing managers must intervene in a system made up of many parts (McGlade and Allen 1985). By ignoring the implications of manipulating individual parts of a system, fisheries management could itself be accused of contributing to the over-exploitation of resources throughout the world. And potentially the situation will continue to worsen, because countries are placing more and more emphasis on their marine resources, e.g. for international trade exchange. These factors, together with the demands put on a country to implement extended jurisdiction under the law of the sea (UNCLOS III) (United Nations 1982), place great emphasis on the acceptance of a systems approach in fisheries management.

One excellent example of such an approach is that of 'adaptive management' – a school started by Holling, Hilborn and Walters at the University of British Columbia (the history of this group and its recent work is given by Walters 1986). Here the premise has been that ecological models *per se* are insufficient to deal with the exigencies of a real harvesting system. Instead management has to adapt actively to new conditions by means of mathematical models and statistics to pinpoint uncertainties and estimate their effects, together with optimization and game-playing procedures to find policies that obviate different problems.

Integrated Fisheries Management operates in a similar direction, i.e. under its aegis, mathematical models involving the concepts of self-regulation, anticipation and feedback are used to generate the long and short-term dynamics of a resource base and its associated industries (Allen and McGlade 1986, 1987a). However, it also represents a further response to some of the spatial and anthropological problems in fisheries. The models are established through numerical and analytical approaches together with simulations calibrated with data from 'real fisheries'. IFM models also attempt to account for as much of the dynamic behaviour of the fishermen as possible by generating details of the spatial dynamics of the resource and industry plus the social and economic effects of various elements such as the global economy, international markets and information exchanges within and between the various components of the industry. Users can therefore explore the long and short-term responses of the system to different management policies and strategies rather than limit their views to a single predictive output. Indeed, it should be stressed that predictions are *not* a substantive output of IFM models, but rather an adjunct to the consideration of a range of possible future scenarios. We believe that the replacement of the 'single prediction' syndrome by a vision of 'probable future states' is not only more realistic when dealing with living systems but also a necessary safeguard in our highly connected and integrated global system.

One of the first steps in the successful implementation of IFM models is to examine the premise that the fishing industry acts as a global regime (Puchala and Hopkins 1982; Majone 1986). By this we mean that the principles, norms, rules and decision-making structures give rise to a convergence of expectations and hence behaviours

(Krasner 1982; Young 1982). Unfortunately, in the case of most fisheries organizations, attention to the details of management has led to the adoption of an institutional perspective, which is merely a reflection of the ability to implement rules and practices. Sadly, then, any convergence of behaviours amongst industry and government is likely to be superficial, with formal and informal aspects operating in different directions. For example, managers tend to have a different perspective from fishermen concerning the long-term goals of conservation, as many fishermen discount the future heavily because of their need to pay off immediate loans and wages. Management is thus often viewed by fishermen as interventionist until a resource is in danger of collapsing, at which time stringent exploitation rules are supported by both managers and fishermen. Finally, behaviours are often biased because fishermen do not want to be involved with management meetings and discussions (Duer Stevenson 1986).

Management problems increase when local concerns are subsumed into larger strategies controlled by national regimes, which rely on specialized agencies for research, information gathering, monitoring and enforcement. Local issues are not so much ignored as 'rolled up' into more expedient packages, for example a demersal fishery in area X, where X can be greater than 500 km^2. Local arguments become even less important internationally, because national representatives on management committees become sovereign in their own right, and must therefore argue for the good of the country rather than for a local community. And yet it has been shown time and again that small-scale events can become critical to the overall operation of a national or international fishery (e.g. blockades of ports by specific gear sectors). Fisheries management must therefore account for the dynamics of the harvested populations, and those of the human populations exploiting them. Yet, to date, fisheries models have dealt almost exclusively with fish population dynamics, treating fishing effort and the behaviour of fishermen and processors as exogenous variables in the calculation of specific harvesting yields. The particular reasons why this situation has arisen are historical and not open to insightful interpretation. What is clear, however, is that studies of the values, norms and beliefs that constitute a way of life in maritime communities around the world provide a serious challenge to the traditional approach of fisheries management (Leap 1977; Orbach 1977; McCay 1978; Acheson 1981; Sinclair 1983; Gatewood 1984; McGlade and Allen 1985; Bailey 1985; Berkes 1985; Glantz 1986; Berkes 1987; Allen and McGlade 1987a). Furthermore ignorance of the socio-economic and political aspects of fisheries can lead to radical changes and sometimes catastrophic situations, as witnessed in the Icelandic and Falkland fisheries 'wars', and the collapse of the Peruvian *anchoveta* fishery (Glantz 1986).

Returning, then, to the idea of a global regime, it is obvious that current behaviours in industries throughout the world may be only superficially conventionalized. For Integrated Fisheries Management models and hence strategies to be successful, it is critical that the underlying rationalities of the various actors in the system be exposed and some attempt made to create a system in which these behaviours can be made to converge.

Integrated Fisheries Management and the marine eco-system

In a recent review by Caddy and Sharp (1986) the role of ecology in marine fisheries was deemed critical to management. However, it must be admitted that defining marine eco-systems and their constituent parts has been an unsuccessful endeavour as regards fisheries. Indeed, what characterizes any eco-system and whether or not it has emergent properties are questions still pondered at length in the theoretical literature. No clear answer currently exists, and it is not obvious that macroscopic features, other than connectivity or energy flows, can be truly discussed with any empirical support. Of even more concern is the evidence, from some theoretical work, that the level of aggregation is in itself critical to the perception of average system behaviour (Horsthemke and Lefever 1984), but the mere process of averaging itself produces a dynamic insufficient to deal with the realities of a complex system. In large part, the models and simulations of eco-systems that exist in the literature ignore these aspects and therefore may never be able to capture the true dynamics of the living systems they hope to describe.

How, then, should the problem of system definition and aggregation be best approached? Traditionally, eco-systems have been formalized through taxa or function. In the Bering Sea model (Laevastu and Larkins 1981) and North Sea model (Anderson and Ursin 1977) organisms are identified taxonomically. In the model of the Benguela Current (Bergh 1986) a mixture of function and taxa is used; for example, squid and hake are identified, but small shoaling fishes and predatory fishes are put together as composites. These three models represent different approaches to the problem of identifying key elements or processes in the eco-system: the Bering Sea model is a compartmentalized accounting model emphasizing information or states, the North Sea model uses simultaneous differential equations emphasizing information on rates, and the Benguela Current model uses a static input–output budget and emphasizes the trophic relationships and the biomass accumulated in each component.

Whichever approach is taken, one overriding problem continues to arise, that is, the spatial and temporal evolution of the eco-system itself. Ricklefs (1987) rightly suggests that the evolutionary context of ecology has been generally lacking. Indeed, without this perspective, questions such as why certain species are found in an eco-system and not others remain unanswered. This has obvious repercussions when asking why certain 'new' species invade an eco-system as a result of intense fishing or over-exploitation, because the answers are highly relevant to commercial fishermen. For example, when fishermen encounter sand lance (sand eel, *Ammodytes*) instead of herring (*Clupea harengus*) or other commercial species, as off the east coast of North America (Sherman 1978), or dogfish (*Squalus acanthias*) instead of haddock (*Melanogrammus aeglefinus*) or cod (*Gadus morhua*), as in the Gulf of Maine (ibid. 1978), entire sectors of a fishery can collapse or be drastically downsized. 'Normal' descriptions of these areas, using functional or taxonomic groups, cannot predict such changes or invasions, because elements outside the system and the factors keeping them out are not examined. Thus, to be of any use in fisheries management, eco-system models must account for the co-evolution and coexistence of different organisms within and between contiguous systems.

Unfortunately, the publication of Beverton and Holt's book (1957) on the dynamics of exploited populations, which supported the way for an ecological view of fisheries, somehow left the impression that dealing with more than one species at a time within a management scheme was not necessary or even critical. Some subsequent analyses have tried to shift the balance in favour of an ecological framework (e.g. Caddy and Gulland 1983; May 1984; Iles 1986; MacCall 1986; Caddy and Sharp 1986), but examples of any biotic interrelationships actually being used to change fisheries management advice are virtually unknown. Even the most documented cases of ecological disturbance and replacement, such as the outburst in the North Sea of gadoids and other commercial species in the wake of a drastic reduction in herring and mackerel (*Scomber scombrus*) stocks due to improved fishing techniques (Daan 1978; Ursin 1982), the replacement of sparids by cephalopods off the coast of West Africa, the collapse of the demersal and pelagic fisheries in the Gulf of Thailand (Pauly 1979) or the replacement in the Peruvian upwelling system of anchovy (*Engraulis ringens*) by other pelagic and demersal species (Arntz and Robles 1980), have not changed general policies or strategies.

The resistance of various fisheries management organizations against the use of even a reasonably restricted multi-species approach is also evidence of the difficulties that lie ahead in trying to formulate

Integrated Fisheries Management models. There have, of course, been several reasons why these approaches have not been readily adopted, not least of which is the fact that they are cumbersome to use, and often the only people to gain any insights about the system are the participants of the modelling process itself. Second, many of the multi-species models tested in the past have been highly sensitive to the functional form of the parameters and initial conditions, so that even the most elegant or 'well thought out' forms have rarely generated realistic behaviours. Indeed, the very diversity and periodic collapse that characterize the components within most eco-systems are not easily mapped by models. Problems that have emerged include insufficiency of prey items, discrepancies in mortality rates for early life history stages and too few cross-linkages. Indeed, the only way of achieving a better understanding of why certain species remain in a system or invade it, is to have a better appreciation of each eco-system's functioning, including the nutrient cycling and inputs from outside sources, the dynamics of the life history strategies of each taxonomic or functional component and the relationship of those species within the eco-system to those in peripheral areas (e.g. Allen and McGlade 1987b).

Integrated Fisheries Management and decision-making processes

From the comments above we conclude that the successful establishment of Integrated Fisheries Management models within the 'policy-making machinery' involves an understanding of the expect-ations and roles of the different actors and components in the fishing industry. Recent work in dynamic systems has shown that the very nature and complexity of real-life situations result from a dialogue between the randomness that exists at a microscopic level and the determinism that may temporarily characterize the average behaviour of a system and its parts. It is therefore crucial that the basis of rational decision-making is well founded on an understanding of the human and biological processes at local, regional, national and international levels. For example, an approach including direct controls on outputs, tax programmes or areal rights impinges not only upon the resource but also upon access to it. This, of course, refers specifically to the issues surrounding common property rights.[1] As fisheries management restricts access to the resource in one way or another, the issue often becomes how best to do it – and it may be that economic rather than biological considerations play the largest role in determining which scenario is fulfilled.

IFM models must therefore be able to address complex questions and provide managers with the ability to choose between different options for action. By stressing the need for a knowledge of the dynamics of self-regulation, anticipation and feedback, these models can help to document which changes in the system are important, and which should be monitored or at least responded to. One of the main aims of this chapter is to present some of these ideas in an existing situation, and so we have chosen a relatively well studied part of the marine environment, the western Scotian Shelf off eastern Canada, where there is a well established demersal fishery of haddock (O'Boyle and Wallace 1986). We hope that by focusing on one area it will become clearer how the approach of IFM can be achieved. In particular we have looked at the economic aspects of the haddock fishery, using a Monte Carlo version of models previously used (see McGlade and Allen 1985; Allen and McGlade 1987a). We explore two possible goals of management: (1) to find a level of fishing effort that 'maximizes' a long-term steady-state yield of haddock and (2) to achieve a harvest of haddock that is steady from year to year.

The first goal is approached by postulating the form of the recruitment and survival functions. The second goal, however, demands that the description of the fishery be dynamic, so as to consider how changes with time affect the responses of the most critical variables. Finally we look at the problems associated with building IFM models in a geographical area that is already well studied by scientists and thoroughly explored by industry, and for which long-term limits on exploitation may have already been exceeded.

THE WESTERN SCOTIAN SHELF MARINE ECO-SYSTEM

Physical environment

The continental shelf in the Gulf of Maine/Scotian Shelf area is part of the continuous Atlantic margin of North America, and represents the southern limit of glaciation of eastern North America. As a result of fluvial and glacial action there is a similarity in the sediment type throughout the area. Sub-surface geology indicates that the Scotian Basin and the Yarmouth Arch are major features that further serve to highlight the continuity of the area.

In the general surface-layer circulation, water moves in a southward direction over the continental shelf inshore of the warm northward-flowing Gulf Stream (Figure 14.1a). The Labrador and Nova Scotia currents bring cold water of low salinity down from the north; as the water moves southwards it warms and becomes more saline but is then

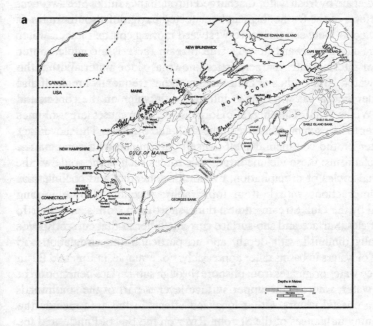

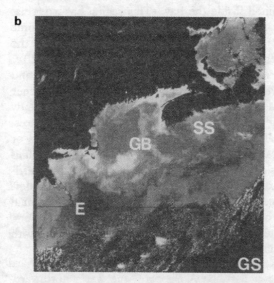

Figure 14.1   *a* The bathymetry and geographical location of south-western Nova Scotia. *b* The same view shown via the coastal colour zone scanner; *SS* Scotian Shelf, *GB* George's Bank, *GS* Gulf Stream, *E* Gulf Stream eddy

diluted again by fresh water discharged through the Gulf of St Lawrence. The effects of the freshening can be detected beyond the southern tip of Nova Scotia. Between the south-westward moving coastal or shelf water and the north-eastward-flowing Gulf Stream there is a complex water mass of intermediate characteristics known as slope water. Within the slope water, large eddies or Gulf Stream eddies frequently break off and meander into areas where they interact with water on the continental shelf (Weibe 1983). The energetic clockwise rotation causes large volumes of water to be exchanged between the shelf and offshore. The movement of water on the continental shelf is itself determined by a combination of mechanisms: these include large-scale forces such as the trade winds, seasonal cycles of precipitation, evaporation, heating and cooling, and tidal interactions with bottom topography; second, offshore forcing caused by the Gulf Stream eddies; third, short-term storms; and, fourth, tidal cycles. Surface and sub-surface currents differ in that current speeds generally diminish with depth, and in particular the circulation and source of waters in basins differ appreciably. For example, in Emerald Basin the deep water originates from offshore through sub-surface penetration of slope water, whereas the upper surface layer is part of the southward-flowing coastal waters. Other localized effects in the area include the freshening influences of the St John River on the Bay of Fundy, and the large tidal range contained inside the bay. (This region has a special shape that causes the tidal wave to 'resonate' and thereby become amplified (Garrett 1972). This, together with the tidal movements in the Gulf of Maine, produces one of the largest tidal ranges in the world.)

Tidal currents in the area have been studied (Garret *et al.* 1978), and the results show that where strong currents occur, such as on Lurcher Shoal off southern Nova Scotia, and in several parts of the Bay of Fundy complete vertical mixing occurs, with important biological consequences. In addition, there are also areas where strong horizontal gradients or fronts occur: one of these is the shelf slope front, which is usually found over the continental slope and marks the point where southward-moving shelf water is juxtaposed to the warmer, more saline slope water. Overall, then, the physical oceanography of the area involves a continual and complex exchange of waters that makes it a highly variable system, and one which generates a weather system that often interferes with fishing activities, especially in the winter months.

Biological environment

The implications of such physical diversity are shown in the production of biological organisms. Phytoplankton biomass increases rapidly in early spring and remains relatively high throughout much of the summer and into the autumn in the Gulf of Maine, but in vertically mixed areas, such as on the Scotian Shelf, the pattern is very much restricted to a cycle that peaks in a few weeks of spring and to a lesser extent in autumn. Measurements of phytoplankton biomass in some vertically well mixed areas often reach values in excess of 2 mg/l (Gordon *et al.* 1980), and the contrast between these and other areas can be clearly seen using remote sensing imagery from the Coastal Zone Colour Scanner (Topliss 1982) (Figure 14.1b). The physical diversity of the area also affects the distribution of the flora and fauna, in that the compression of many ranges generally makes them more difficult to analyse compared to equivalent latitudinal regions in the eastern Atlantic, where environmental changes are gradual.

The floral and faunal assemblages in the area are essentially northern and boreal. On most offshore banks there are no macro-algae attached to the sea bed; instead there are micro-algae ranging in size from small net phytoplankton (0.02 mm to 0.1 mm), through nanoplankton (0.001 mm to 0.02 mm), to picoplankton (less than 0.001 mm). The dominant species of the area are representative of a boreal assemblage, e.g. *Ceratium* (O'Reilly and Evans-Zetlin 1982). The ratios of the various size assemblages vary seasonally (e.g. off Yarmouth net plankton dominate during the winter months but not in the summer), and the data suggest that these characteristics are associated with local oceanographic features and processes and season, rather than with a larger-scale geographic trend.

The dominant zooplankton are also members of the boreal assemblages (e.g. *Calanus finmarchius*, *Metridia lucens*, *Temora longicornis* and *Pseudocalanus*) (Fleminger and Hulseman 1977; Cox and Wiebe 1979). There are also representatives of Arctic species (e.g. *Calanus glacialis* and *Calanus hyberboreus*). Slightly to the south a transition area between zooplankton of northern and southern origins has been identified (Haydon and Dolan 1976).

The benthic organisms in the area and their distribution reflect substrate distributions, in particular the muds, coarse sand and gravel (Wigley 1968; Peer *et al.* 1980). Species such as horse mussel (*Modiolus modiolus*) exist in dense beds, as do sea scallops (*Placopecten megallanicus*) in areas of similar substrates. Macrobenthos are less abundant on both sandy and silty substrates. The seventy-nine different taxa of benthic invertebrates (Wigley 1968) listed from the area are generally characteristic of the Nova Scotian biogeographic province.

Detailed consideration of the fishes and invertebrates of the area gives at least eighteen species of commercial importance (Mahon *et al.*

1984). These can be categorized into two groups: widely distributed (shortfin squid *Illex illecebrosus*, lobster *Homarus americanus*, sea scallops, Atlantic herring, Atlantic mackerel, white hake *Urophycis tenuis*, silver hake *Merluccius bilinearis*, swordfish *Xiphias gladius*, bluefin tuna *Thunnus thynnus* and yellowtail flounder *Limanda ferruginea*) and northern species (American plaice *Hippoglossoides platessoides*, argentine *Argentina silus*, cod, cusk *Brosme brosme*, haddock, halibut *Hippoglossus hippoglossus*, pollock *Pollachius virens* and redfish *Sebastes* sp.) (Halliday *et al.* 1986). Computations of the annual production of fish indicate that the area is comparable to areas in the Gulf of Maine and farther south in the mid-Atlantic Bight (Mills 1980).

Whilst no ecological model has been put together for the area off south-western Nova Scotia, recent initiatives on Brown's Bank and George's Bank by the Department of Fisheries and Oceans (Canada) and the National Marine Fisheries Service (USA) have focused attention on the possibility of defining ecological regimes in the ocean. Unfortunately the evidence has been inconclusive and emphasizes the fact that, despite the large amount of time and effort spent examining the biological aspects of the areas, a functional ecological model of the area is still lacking. In fact detailed studies of the population dynamics of single species are only now being undertaken. For example, data on diet changes in haddock support the idea that the species undergoes a transition from a pelagic to demersal phase (Mahon and Neilson 1987); the authors concluded that haddock juveniles live and feed in the epipelagic zone until they are ready to assume a demersal existence, at which point the fish make forays to the bottom until a suitable substrate containing benthic prey has been located. The transition may be instantaneous for individuals, but occurs over a month for the cohort (Koeller *et al.* 1986). Understanding these dynamics is critical to our knowledge of the functioning of the eco-system and of the likelihood that recruitment of a species such as haddock will undergo fluctuations. But progress in this area is limited by the amount and extent of resources available for research: the recent suggestion that large-scale eco-system models (LSE) are necessary for fisheries management (Sherman 1986) should thus be carefully scrutinized. It is not enough to measure 'everything all the time', because, as has been stated, what is often critical is what lies around the edges of a system, not what is inside. Moreover the timing of localized events could never be monitored for an entire eco-system. Rather, process-oriented studies will have to be taken as a proxy for events throughout the system.

In the Scotian Shelf system as a whole there are a number of commercially important species of fish whose population dynamics are thought to have been affected in some way by fishing. The question is what these effects are, and how important they are for the long and short-term viability of the resource and the fishing industry in the area. To understand the response of a species such as haddock to exploitation we must be able to describe and quantify the dynamic nature of any life-history changes, the form of the recruitment function and the impact of both these aspects on the sustainable development of the fishery. To measure everything in the system will not help to answer these questions, because the processes involved will have economies of scale and a certain granularity; for example, coarse-grained aspects could be related to oceanographic and weather conditions, whilst fine-grained effects could range from the spatial juxtaposition of fish and fleets to that of fish and plankton. In other words, the key functional forms and points of interaction have first to be defined and then used as a basis for building IFM models. In a way this step may simply be an exercise in common sense; however, it more often exposes the *ad hoc* nature of models and policies, plus shortcomings in many biological data bases available at present for fisheries management.

THE ECONOMY AND CULTURE OF SOUTH-WEST NOVA SCOTIA

In south-west Nova Scotia there are no large cities or towns and no strong service sectors (Figure 14.2a). Fishing and mining are the two main employment contributors to the gross product of Nova Scotia as a whole. In some years mining contributes more, but there are no significant mining developments in south-west Nova Scotia – the area that we are most concerned with in the model. Agriculture is only marginal, because of poor soils in the area. Forestry, although important, lags far behind fishing in regional importance. There are approximately seventeen ports of note in Yarmouth, Shelburne, Digby, Annapolis, Queen's and Lunenburg counties; landings are made here from the offshore and inshore fisheries. The most important fish and invertebrate species harvested by these fisheries are scallops, lobster, cod, haddock, white hake, pollock, redfish, plaice and winter flounder. The principal cash species are scallops and lobsters, followed by groundfish species such as cod and haddock.

Fish processing is the largest activity behind the primary harvesting in the province, accounting for at least fifteen per cent of total manufacturing employment – a level almost twice that of the pulp and paper industry. The provincial economy is heavily dependent upon foreign markets, and fish products are the major export of the province.

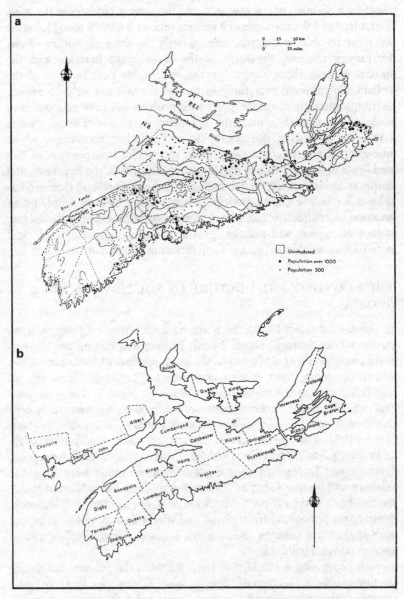

Figure 14.2 *a* The population distribution of Nova Scotia. *b* The counties of Nova Scotia

As with the fishing activity in the province, seventy-five per cent of the provincial processing industry is situated in south-west Nova Scotia. Unfortunately, the processing plants are highly dependent upon access to and demand from markets; the impact of tariffs and reduced winter transport links across the Gulf of Maine, for example, all serve to frustrate the export of goods and highlight the critical nature of outside markets for the area.

Outside the industrial-scale plants there are many small facilities scattered along the coastline in villages that are characterized by a rural economy totally dedicated to the fishery. These villages depend upon the fish plants and fish for their survival and social structure, because the economy of the area is so undiversified. The fishing industry also requires goods and services, and so throughout south-west Nova Scotia suppliers of the industry are also an important feature of the economy. They comprise net and gear manufacturers, boatbuilders, electronic gear manufacturers, etc. These account for approximately five per cent of the total area employment. In summary, then, the resource-based and largely rural economy of south-west Nova Scotia is driven and sustained by the wealth created by exploiting the marine eco-system. In this respect the area is very similar to Iceland and Norway. In fact the proportion of employment generated by the fishing industry in those countries is even lower than in Nova Scotia.

The social geography of the area serves to emphasize the points made above: the road network skirts the coastline, leaving a hinterland almost devoid of connecting routes. In many instances the labour force can be said to be mobile, with trawlermen and plant workers moving in the region according to employment opportunities, but with families generally staying in one community over time. This pattern lends stability and cohesion to many of the small villages with no obvious link to any particular fishery.

The ethnic background is somewhat mixed: in the counties of Queen's and Lunenburg a large segment of the population is of German origin; in Yarmouth and Digby about forty per cent of the population are Acadians of French origin; and in Shelburne the vast majority are of British origin (Figure 14.2b). Given this diversity, it is interesting that the variety of species sought for the regional markets is extremely limited compared to the range available in the eco-system itself. It is these features that we have tried to capture in previous models, by placing a higher demand for certain species in certain areas (Allen and McGlade 1986, 1987a).

THE HADDOCK FISHERY OFF SOUTH-WEST NOVA SCOTIA

Background information

The haddock fishery off south-west Nova Scotia considered in this chapter is managed within the Northwest Atlantic Fishery Organization (NAFO) Division 4X (Figure 14.3), although the resource itself is part of a continuum from sub-area 5 and partly determined by spawning activity in Division 5Zj (Figure 14.4) (Clark *et al.* 1982). In total dollars of fishing revenue from demersal species in Division 4X, haddock is worth approximately the same as cod, and significantly more than pollack, white hake or flounder. However, the haddock fishery differs from the others in two respects: first, the price of haddock is higher, so

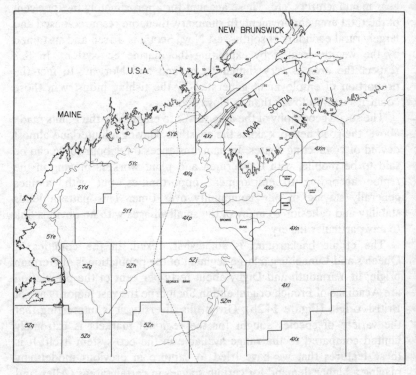

Figure 14.3 The Northwest Atlantic Fisheries Organization divisions and subdivisions

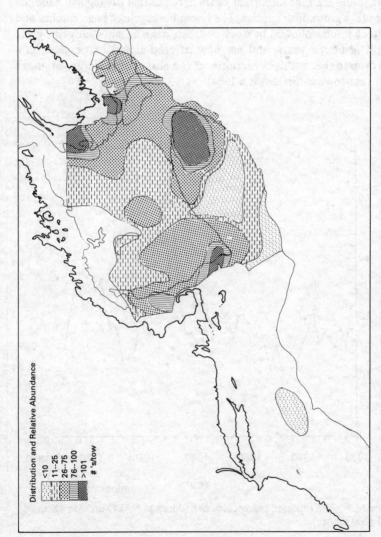

Figure 14.4 The distribution of haddock off south-west Nova Scotia and in the Gulf of Maine, derived from Canadian research surveys (numbers per tow, 1978–84)

Distribution and Relative Abundance

<10
11–25
26–75
76–100
>101
's/tow

it is more vigorously sought after by fishermen and processors, and second, as discussed below, the dynamics of the fishery are more extreme and pronounced.

At present, haddock abundance is depressed (Figure 14.5 and Table 14.1) (O'Boyle and Wallace 1986); indeed harvests have not been as low since 1972–74. The average catch per year since 1930 has been 20 000 tonnes, whereas in 1985 it was 15 000 tonnes. The periodic oscillations of large amplitude that are observed in the catch, fish stock size and effort levels are thus important characteristics that distinguish haddock from other groundfish species. The boom/bust cycle of huge catches and fishstock levels, followed by stock collapse, has a nearly steady one-cycle period of fifteen years, and has now covered almost two cycles. As a backdrop to this, haddock recruitment can also undergo large variations from year to year (Sissenwine 1984).

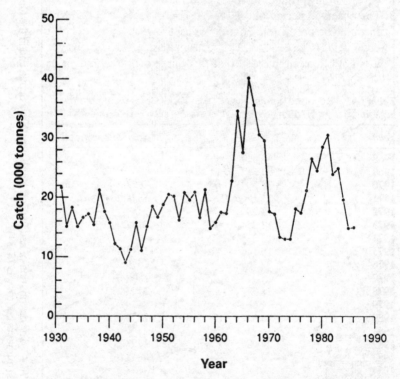

Figure 14.5 The reported yearly catch of haddock in NAFO division 4X since 1931 (000 tonnes)

The first question, then, is why the fishery exhibits an oscillatory state, particularly with a period of fifteen years, when variation in recruitment is determined annually. A clear answer to this question would not only help us to construct a model which could be used to stabilize the supply of this resource, but would also help to create a conventionalized set of expectations and hence behaviours in the demersal fishery.

One explanation is that the fishery is undergoing predator–prey (Lotka–Volterra) oscillations (Schaefer 1954; McGlade and Allen 1985; Allen and McGlade 1986, 1987a). Five observations support the idea that haddock are acting as a prey item with the fishery as predator:

1 All three observed variables, i.e. catch, fishstock and fleet size oscillate with the same period of fourteen – fifteen years.

2 The motion represented as a trajectory on an effort/fishstock size diagram is counterclockwise, with the fish leading the vessels through the cycle by about 90° (see Allen and McGlade 1986: Figure4). This is true of any predator–prey system.

3 The observed period of oscillation, $T = 15$ years, is in agreement with the analytic prediction from the Schaefer–Gordon model of fisheries economics (Clark 1985). Although long understood (Schaefer 1954), it is not often emphasized that this fundamental model, upon which

Table 14.1 Reported nominal catch (*t*) of haddock (*Melanogrammus aeglefinus*) from NAFO Division 4X, excluding sub-division unit 4Xs (tonnes)

Year	Catch	Total allowable catch
1968	30 622	–
1969	29 612	–
1970	17 582	18 000
1971	17 166	18 000
1972	13 281	9 000
1973	12 954	9 000
1974	13 035	–
1975	18 144	15 000
1976	17 365	15 000
1977	21 217	15 000
1978	26 577	21 500
1979	24 631	26 000
1980	28 612	28 000
1981	30 746	27 850
1982	24 087	32 000
1983	24 990	32 000
1984	19 681	32 000
1985	14 927	15 000
1986	15 034	15 000

fisheries economics is largely based (Anderson 1986), is a theory that predicts predator–prey oscillations about the open-access steady state. In fact a broad range of fisheries economics models which include Schaefer–Gordon as the simplest example also predict predator–prey cycles for the 'standard' one-species fishery. The assumptions needed for an oscillatory fishery will be discussed below and are employed in the simulation model used in this chapter. Solving the simplest two-variable form of the Schaefer–Gordon model (Schaefer 1954; Smith 1969; Clark 1976) can be carried out analytically, yielding a prediction for the period of one cycle. The theoretically predicted value is Tp = 18 years. Such close agreement, eighteen years predicted, compared to fifteen as observed, is an encouraging validation of the model.

4 There is, between model and reality, a good correspondence in their qualitative features as well. Both models and fishery cycles are characterized by a period that is relatively steady, while the amplitudes in both model and reality vary considerably.

5 It is necessary, however, to investigate other possible sources of cyclicity. If the environment underwent cyclical variation due to climate or the tides, for example, conditions for fish reproduction and survival could vary in the same manner, giving rise to a cyclical fishery. The evidence seems to indicate that this is not the situation with Division 4X haddock. Looking at the catch time series of Figure 14.5, there is no sign of an oscillation before 1960, the one large drop from 1940 to 1945 being the result of World War II. Environmental factors normally act over long time-scales; for example, it is impossible to imagine the tidal cycle or a sunspot cycle having suddenly begun in 1960. On the other hand, fishing effort increased substantially during the 1960s, especially by Canadian offshore and foreign vessels. The sudden onset of the cycle at a time when effort was rapidly increasing strengthens the predator–prey hypothesis and diminishes the probability of environmental forcing as a primary cause.

It is also worth noting from previous studies (Allen and McGlade 1987a) that cod and pollock exhibit no oscillatory behaviour. This may be for possibly two reasons: first, fishermen do not respond predictably to cod or pollock stock size because of the lower prices that these two species bring in the market place, and, second, recruitment of cod and pollock does not vary as greatly as haddock from year to year.

The model: initial description

Both spatial and non-spatial models have previously been developed for this fishery (Allen and McGlade 1986, 1987a); the model described below incorporates Monte Carlo techniques in an attempt to improve on the earlier versions, and to allow further exploration of different management strategies.

The simulation model incorporates the same features that characterize any predator–prey fishery. It hypothesizes a stock–recruitment relationship, a catch proportional to both vessel number and fish stock abundance, and a level of effort which changes at a rate proportional to profit. The variables are fish stock, F, and number of boats, B. The real yearly price of haddock paid to fishermen has remained relatively stable over the period to be modelled, so it is taken as fixed. Cost is set proportional to boat number. The fish stock is divided into three cohorts, ages 0 to 2, 2 to 4 and 4+. A logistic stock–recruitment relationship is employed which incorporates a stochastically varying parameter b, described below, which simulates yearly recruitment variation.

The parameters of the simulation model are evaluated by assuming that the steady state (F_{ss}, B_{ss}) at which $dF/dt = dB/dt = 0$ can be approximated by the average (F, B) of $\{F(t_i)\}$ and $\{B(t_i)\}$ taken from years $\{i\}$ spanning one complete period of the oscillation. The values of the two yearly time series of fishstock and effort $\{F(t_i)\}$ and $\{B(t_i)\}$ are taken from the date for $t_i = 1970$ to 1984. This period is the most recent, and the data collection processes were better then than in earlier periods. It should be noted that the fluctuations in amplitude of the predator–prey cycle could affect the accuracy of this assumption because the time-series averages would differ from the steady-state values. Apart from this, the general nature of Lotka–Volterra systems in which $(F_{ss}, B_{ss}) = (F, B)$ implies that this is a very good approximation.

Since this simulation model is constructed to find policies which lessen the bust/boom highs and lows, all possible factors which might be enhancing the cycle should be included if possible. It is these destabilizing factors that a management policy must seek to overcome. May (1974), Maynard Smith (1974) and others have shown that discreteness and stochasticity are both destabilizing and thus exacerbate predator–prey oscillations. Therefore processes which are best represented as discrete or stochastic are modelled as such; specifically, the variability of recruitment and the discrete and time-stochastic nature of catch and entry/exit are incorporated into the model.

The number of boats in a fishery is an integer. When a boat enters, the change happens at a discrete moment in continuous time, t' and the variable B goes from $B(t')$ to $B(t') + 1$. Once an entry/exit event occurs, the amount of waiting time until the next arrival or departure is not a predetermined quantity but can vary stochastically, and certainly does in a real fishery, since the two events are normally unrelated. What matters for the macro-scale dynamics is that the *average* number of entry events per year increases when the fishery is profitable. The method of birth and death simulation (Pielou 1969; Gillespie 1977) is ideal for modelling this process, since it simulates a discrete change of boat number at a discrete moment in time. The waiting time between events is chosen from a probability distribution which is derived assuming only that consecutive events are independent and occur with an average entry probability that increases with profit. The catch process is also modelled with this technique, where the discrete event being simulated is the landing of each fishing trip.

One particular advantage of the birth and death method is that it serves to bridge the separation in understanding between the microscopic or fine-grained and macroscopic or coarse-grained levels of interaction. Each time a simulated boat enters or leaves the fishery it does so explicitly at a stochastically chosen time. Likewise, each individual catch event, namely a fishing trip, is simulated and a catch reported. By summing up the effect of many micro-events the macro-scale time evolution is modelled.

Normally an understanding of events on the micro-scale is better because we can observe them direct. The modeller can go out on a fishing trip himself or can interview a fisherman if he goes out of business. To examine the macro-behaviour of an entire fishery one must analyse large sums of numbers. In this statistical form, intuitive understanding is limited. Birth and death methods as used in this chapter allow a modeller to test observations about micro-scale events, such as a fishing trip, to see whether allowing these to occur, for example, twenty times per year per boat at stochastically chosen times gives rise to the observed macro-behaviour that the fishery statistics report. When the simulation does agree with reality, it strengthens and validates understanding of the fishery at both levels.

The effect of a stochastically varying recruitment coefficient was first studied by Nisbet and Gurney (1982) and applied to fisheries by McGlade and Allen (1985). The parameter used for achieving this, b, is simply the ratio of recruitment numbers divided by parent stock biomass calculated each year. When b is constant, implying that the fish stock strictly obeys a logistic stock recruitment formula, the steady state is stable and oscillations will occur but will quickly die down. When b

Table 14.2 Estimates of the coefficient *b*, for haddock (*Melanogrammus aeglefinus*) in NAFO Division 4X

Year	b
1962	2.14
1963	4.95
1964	13.55
1965	0.49
1066	0.34
1967	2.00
1968	0.37
1969	1.16
1970	5.47
1971	0.20
1972	17.50
1973	16.31
1974	5.38
1975	15.44
1976	7.49
1977	2.01
1978	4.53
1979	4.07
1980	4.49

varies in a probalistic (i.e. stochastic) manner, the amplitude increases and the oscillation continues indefinitely on its irregular counter-clockwise motion, as observed in Division 4X haddock.

Since variation in recruitment plays a vital role in causing the pathology of bust and boom, it is important to model this yearly stochastic variation as well as possible. Rather than attempt to fit the data to some previously known distribution, we incorporated stochastic time-series information direct into the simulations. First, the range of observed values of *b* is partitioned into a set of sub-intervals. The manner in which this is accomplished is described below, using as an example values of *b* derived from O'Boyle *et al.* (1984) and given in Table 14.2. This time series of *b* yields a range of nineteen values from 0.2 to 17.5. We evenly subdivide the range into eight sub-intervals. The frequencies of *b*s falling in each sub-interval are given, as are the normalized probabilities, in Table 14.3.

Stochastic recruitment is modelled by sampling from this distribution of *b* each simulation year in the following manner. First, the random number generator selects a number R from 0 to 1. If $R < 0.4$, the first sub-interval is chosen; if $0.4 < R < 0.6$, *b* is chosen from the second subinterval, and so on. Since the random number generator chooses the

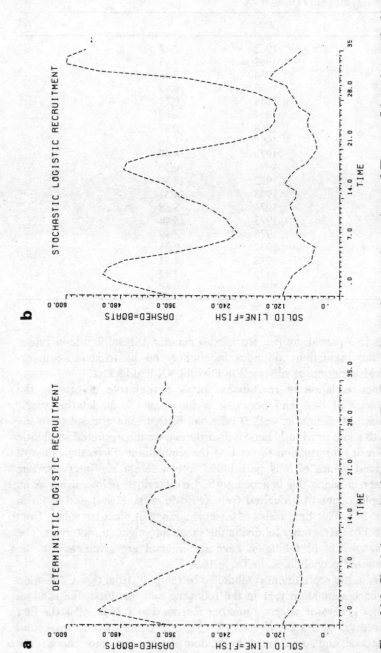

Figure 14.6 Simulation results of the model. *a* The fishery with a perfectly steady recruitment. *b* The same fishery characterized by a recruitment which varies stochastically from year to year.

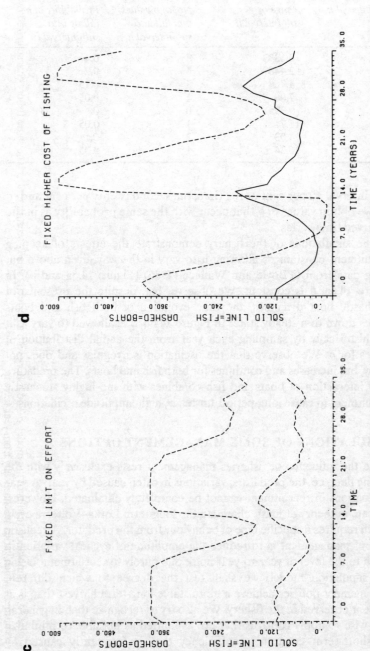

c The effect of a fixed upper limit on boat numbers at the steady-state level of 380 boats. d The effect of a fixed increase in the yearly cost of a fishing licence from $78 000 to $92 000

Table 14.3 The range of *b* values by sub-interval

Sub-interval i:	Range of sub-interval i:	No. of b values that fall into sub-interval i:	Probability of b falling into sub-interval i:
1	0.1333	8	0.4
2	2.448	4	0.2
3	4.764	3	0.15
4	7.08	1	0.05
5	9.395	0	0
6	11.71	1	0.05
7	14.03	2	0.1
8	16.34	1	0.05

first interval 40 per cent of the time, the second twenty per cent and so on, we obtain values of *b* that occur with the same probability as in the observed time series.

The simulations of the fishery demonstrate the effects of keeping recruitment constant or allowing it to vary in the way given above but using data from O'Boyle and Wallace (1986) (Figure 14.6a and b). In Figure 14.6a, *b* is constant. We observe that, despite the presence of stochastic boat dynamics, the time evolution of both fish and boats settles down to a steady state. In Figure 14.6b, *b* is allowed to vary as it has historically by sampling each year from the actual distribution of values for *b*. We observe that the oscillation is irregular and does not decay, but increases and continues for both fish and boats. The predator–prey interaction of boats and fish combines with the highly stochastic recruitment to cause long-period, unsteady, high-amplitude oscillations.

SIMULATIONS OF SOME MANAGEMENT OPTIONS

Since the influence of fisheries management rests exclusively with the fishing fleet, i.e. the predators, variations in catch caused by year-to-year fluctuations in recruitment cannot be completely eliminated. However, we can, in theory at least, eliminate the long-term Lotka–Volterra cycle which requires a specific type of behaviour from the predator. A realistic goal of management is to reduce the amplitude of cyclical variation in catch to the level of year-to-year noise due purely to recruitment. Using the simulation model, we shall test the degree to which different management policies achieve a more stable long-term harvest than is at present observed in the fishery. We also try to recognize that, in general, it is wise to apply controls that are as limited as possible to minimize the short-term economic inefficiency that is frequently caused by management restrictions. This can often be achieved by applying controls to the fishery which are designed to work at the correct time. In our examples, fishing will be variously encouraged or discouraged by altering the cost in such as way as to 'damp down' the oscillations in catch.

Four modes of management control have been examined. First, a fixed limit on boat number. Second, a fixed increase in the cost of fishing, which may be imposed by fishery management through the price of a fishing licence. Third, time-dependent controls which serve to increase or decrease the cost of fishing proportional to changes in the profits to fishermen. Fourth, time-dependent changes in yearly fishing costs that are set in proportion to changes in fish stock biomass.

Fixing maximum effort at the open access steady state (Figure 14.6c) induces a minimal improvement in resource stability, failing to reduce the depth and frequency of the 'busts' while eliminating entirely the would-be 'booms'. However, lowering the cap on effort below the open access level achieves both fundamental objectives of fisheries management: stability and optimum net economic yield. The latter objective of a profit from the fishery and this means of achieving it, namely lowering fixed effort to reduce overfishing, are now virtual folk remedies of fisheries science. Based on the equilibrium surplus production model of Gordon (1954) and Schaefer (1957), the strategy of lowering effort increases sustainable yield by reducing intraspecific competition, thus increasing the intrinsic rate of stock population growth. If effort is reduced further (depending in each fishery on fishing cost and landed price of fish), costs decrease faster than revenues, and large profits are possible for fishermen who remain. Simulations similar in form (but not illustrated), left effort free to vary (with profit) below specific maximum limits of boat license number, but when effort reached this fixed cap, management would allow no new vessels into the fishery. The implication of these simulations is that the most robust and reliable prescription for obtaining the dynamical objective of a *stable* yield is the same one which maximizes equilibrium rents: effort held at levels well below open access equilibrium is also the best way to achieve stability.

In the next simulation (Figure 14.6d), the effect of a fixed increase in the yearly cost of a fishing licence is examined. (The initial value of $78 000 that was used in the previous simulations was increased to $92 000, because in trial runs this value was the lowest that produced a significant change in the amplitude of the cycle). The oscillation amplitude of boat numbers becomes very large and the fishery unstable.

The second set of simulations look at time-dependent controls. Hypothetical fishery managers adjust the incentive to enter or leave each year by fixing a variable cost for a fishing licence. Of critical import-

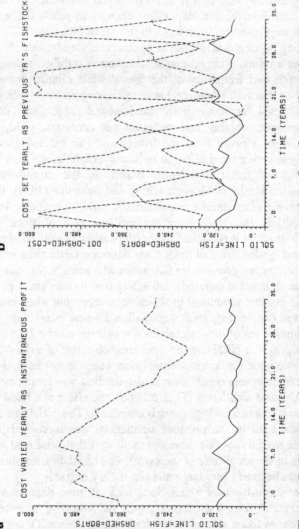

Figure 14.7 Simulations of time-dependent management strategies. *a* Cost of entry adjusted instantaneously proportional to yearly profit. *b* The conventional strategy: incentives that encourage fishing when the fish stock is abundant and discourage fishing when the stock is depleted.

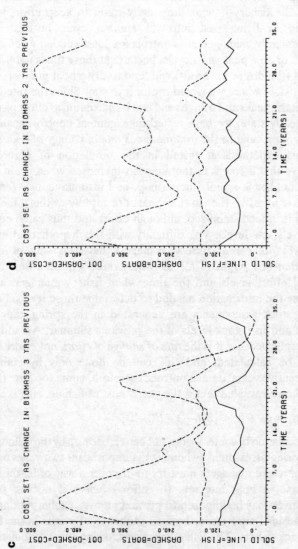

c Cost controls determined by the rate of change of the fish stock three years earlier. **d** A successful management criterion: the cost of a fishing licence set proportional to the rate of change of the fish stock two years earlier

ance in stabilizing the oscillation is the choice of when to encourage or discourage fishing in relation to where the fishery is along its fifteen-year cycle. Two observations borrowed from the mechanics of oscillations are helpful. First, damping of any sort, such as friction, is resistance to velocity. The analogue to velocity in a fishery is the rate of change of the variables, dF/dt and dB/dt. Management can only affect fishermen. Since the goal is to increase 'friction', an effective management policy should attempt to slow dB/dt. Second, resistance to change should be strongest when the fishery is near the steady state, to keep effort levels there, instead of flying past into yet another cycle of bust and boom. Conveniently, management controls are least needed when effort is in a trough or on a peak of its cycle, because at those times natural economic forces (the desire for profit) will tend to drive boat numbers back to the steady state where we would prefer it to stay. Since the rate of change is zero at the peaks and troughs and is at its maximum as effort passes through the steady state, we have tested management control incentives which are proportional to the *instantaneous rate of change of boat number*.

This hypothesis is tested in the simulation of Figure14.7a. We observe that if cost is instantaneously increased when profit is high (i.e. boat number is rapidly increasing), and instantaneously reduced when profit is negative (i.e. so that boats are rapidly exiting), then the oscillation is indeed stabilized, although effort and thus catch equilibrate at relatively low levels. The difficulty with this hypothetical management policy is that real fishery managers have no way of determining *instantaneous profit*. In a real fishery there will be a time lag between actual effort levels and the time when fishery managers receive and analyse the information needed to determine those levels. For example, when stock assessments are generated in the spring, they report the effort and fish stock levels of the previous summer. An additional time lag results because it is the *rate of change of effort*, not effort itself, which must be calculated, and this can be done only by comparing two consecutive years' data. Controls, therefore, come too late. (Simulations not shown revealed that if the yearly cost of fishing, $c(t_i)$ is set to be:

$$c(t_i) = c_0 + h \times [B(t_i-1) - B(t_i-2)]$$

the oscillation becomes worse, not better, because by the following summer the spring assessment (when $c(t_i)$ is imposed) is two years out of date.)

Therefore managers must try to discover a way of learning about the increase in boat numbers (or effort) *before* it occurs, so that cost incentives may be imposed to prevent an over-exuberant fleet response.

Consider once again the time sequence of events in the Division 4X haddock fishery. The fish stock grows, owing to a combination of good recruitment and lower effort, then catches increase with the improved stock levels, and finally boat numbers rise owing to reports of a good harvest. Since the rise in fish stock happens first, we can predict the increase in fishing effort beforehand by monitoring the level of fish stock. In Figure 14.7b the cost is set proportional to the previous year's absolute level of fish stock; in particular, the cost of fishing is set higher when the fish stock is greater than the mean, F, and is reduced when the stock is depleted.

$$c(t_i + 1) = c_0 + h \times [F(t_i) - F]$$

The oscillation becomes wildly unstable. This is a policy that managers often employ in real fishery situations, and these preliminary results indicate that in a cyclical fishery like Division 4X haddock it could be extremely unwise.

A better policy would be to set cost proportional to the rate of change of the fish stock. Qualitative analysis of time-series data indicate that the lag between dF/dt and dB/dt is between three and two years, which inspired the simulations of Figure 14.7c and d respectively. The policy which alters cost proportional to the change in fish biomass three years before (Figure 14.7c) essentially fails. The bust/boom cycle is barely affected and the number of boats collapses. In contrast, the management policy employing a two-year lag between the rate of change of biomass and the imposition of cost controls (Figure 14.7d) works reasonably well. By looking at two years and dividing by 2, in order to smooth out recruitment fluctuations which distort the rate of change in fish biomass, and by calculating the cost as:

$$c(t_i) = c_0 \times \{1 + [0.3 \times [F(t_i-1) - F(t_i-3)]) / (2 \times 40)]\}$$

where 40 = the maximum two-year change of fish stock biomass, $[F(t_i-1) - F(t_i-3)]$ observed in the time series, $0.3 = 30$ per cent variation, and c is the original uncontrolled cost per boat per year. This supposes that the fishing industry would in reality not tolerate a management scheme that altered its costs by more than 30 per cent in a year. Particularly encouraging and surprising is that in addition to avoiding any sign of a cyclical fish-stock and effort collapse, this two-year policy results in considerably greater long-term yields, since fish stock and effort remain at relatively high levels. This result would be very sensitive to the rates at which fish stock and boat numbers change in respect of one another. If, in a different fishery, boat numbers increased faster than the two to three-year lag observed in Division 4X haddock, a different cost policy employing a shorter lag would probably prove more effective. Such a relationship between cyclical behaviour and the lag required to alter it is similar to generic results obtained by Huberman and Hogg (1988) for open systems.

CONCLUSIONS

The simulation results presented here lead us to conclude that it is possible to build IFM models that can be used for management decisions. From these we can begin to judge whether or not certain effort controls are effective in maintaining a fishery, and whether indeed there are limits to exploitation. For haddock, the largest unanswered question concerns the causality of fluctuations in recruitment. Given that fishing on this resource can itself exacerbate oscillations, are there any critical periods in the long-term cycle of change in recruitment that are more susceptible to changes in anthropological or environmental factors (e.g. offshore Gulf Stream gyre activity; McGlade 1987)? Evidence from many sources supports the idea that haddock in Division 4X have a shorter spawning period compared to other gadoids; given the dynamic nature of the oceanographic features around south-west Nova Scotia, it is conceivable that variability in recruitment could occur on an annual basis as a result of differences in ocean conditions. However, it must be remembered that the Lotka–Volterra dynamics produce a result very close to the long-term cycle of recruitment of haddock, suggesting that the most important aspects of the exploitation of this species are not so much the inter- and intra-specific interactions but rather those of the fishing fleets.

In this regard, the economic background of south-west Nova Scotia is an important element in any discussions about management, because it is clear that there is little other than the fishing industry to support the various coastal communities. Fishermen in south-west Nova Scotia may be less willing to respond to controls on access, compared to fishermen in areas where viable alternative employment opportunities exist. Indeed, the experience of fisheries management in Division 4X indicates that the vision of the resource among fishermen in the area is somewhat short-term and rather more optimistic than the biological indications suggest (O'Boyle and Wallace 1986). The differences in perception of the resource are the cornerstone of much of the conflict in the area, and emphasize the need for IFM models to incorporate the appropriate responses to controls put in place by the government. It is not simply a case of placing limits on catch and then relying on enforcement and biological models to find the truants and impacts of overfishing. Indeed, the simulations represented here provide a very interesting insight into types of controls that could be considered appropriate on paper but doomed to failure in practice. We would suggest instead that a full array of controls be tested for the haddock fishery in Division 4X, including those on recruitment to the fishing gear (i.e. mesh size), effort limitations and catch controls, using a series of models that incorporate the dynamics of the species and fleets, plus the economic losses that communities are willing to suffer, given a predilection for fishing and the ability to remain in the coastal areas of Nova Scotia.

In more general terms we should also like to return to a point raised earlier. It is because the impact of man as an apex predator is so disproportionate to its size that the large-scale eco-system approach, suggested by Sherman (1986) and others, is a non-viable alternative for fisheries management. This is mainly because the effects of the fishing industry on the biology are felt most acutely at the higher trophic levels, so that information concerning lower levels, whilst potentially important for long-term trends in overall biomass, is of little strategic value to the fishing industry and managers. This is not to say that such data are of no use; rather, the spatial and temporal distribution of resources in the *same trophic level* (e.g. the replacement of one species of fish by another as a result of overfishing) may be of far more importance to the deployment of local and global fleets.

Finally, we would conclude from these simulations and previous work that the fisheries resources of the world cannot be left to the benign neglect of the current biological models of fisheries management which treat the anthropology of fishing as an externality. Detailed analyses of the fishing industry, its economics (whether capitalist or some other type of political and financial regime), spatial deployment and response to controls must be explicitly included in the decision-making that goes into setting catch quotas and effort controls. Indeed, the flexibility of IFM models is necessary in complex multi-objective situations, as for example when artisanal and highly industrialized fisheries come into direct conflict.

Overall, then, we believe that in the future resource management must proceed along a more highly integrated path, because, as shown in the example of the haddock fishery off south-west Nova Scotia, biological and socio-economic factors can have a critical impact on both the short and the long-term viability of the industry. Three more general conclusions are:

1 The oscillatory behaviour of a fishery can be exacerbated by the socio-economic background of the communities involved.
2 Large-scale eco-system approaches to fisheries management are likely to be non-viable because of the scale and dynamic nature of the impact of human intervention.
3 As countries place their resources under more pressure, fisheries management is likely to fail unless there is a concerted effort to

understand and rationalize the behaviour of fishermen, processors and managers within a unified regime.

APPENDIX STOCHASTIC SIMULATION OF THE FISHING INDUSTRY

The birth and death method employed here simulates the master equation (i.e. a discrete Markov process). Birth and death is, in fact, the simplest and most widely studied form of master equation process. We now describe the simulation method for a hybrid system, that is to say, a system whose variables are changing owing to continuous deterministic processes and birth and death events at the same time.

In order to construct the birth and death part of the simulation we must choose the set of N_E events, $\{E_1, E_2, \ldots E_{NE}\}$ which happen probabilistically at discrete times. Each occurs with an average frequency or rate of $\{A_1, A_2 \ldots A_{NE}\}$. And we must define the effect induced by each event. An event can change one or many of the variables. The set of effects is $\{\Delta \vec{X}_1, \Delta \vec{X}_2, \ldots \Delta \vec{X}_{NE}\}$ where the change vector associated with event 1 for example is

$$\Delta \vec{X}_I = (\Delta X_{11}, \Delta X_{12} \ldots \Delta X_{1N_v})$$

where N_v is the number of variables. Owing to event 1, therefore, the change in each variable is:

$$X_1 \rightarrow X_1 + \Delta X_{11}$$
$$X_2 \rightarrow X_2 + \Delta X_{12} \tag{14.1}$$
$$X_{N_v} \rightarrow X_{N_v} + \Delta X_{1N_v}$$

The continuous evolution is described by the usual system of equations.

$$dX_1/dt = f_1(\vec{X})$$
$$dX_2/dt = f_2(\vec{X}) \tag{14.2}$$
$$dX_{N_v}/dt = f_{N_v}(\vec{X})$$

Included in (14.2) are the parameters which may vary stochastically. Here we shall assume that only the parameter b, representing the intrinsic birth rate, can vary, and we shall show how to select b each year from a Gaussian distribution (Abramowitz and Stegun 1970), given the mean and the standard deviation. We will record (i.e. PRINT) the values of the variables after every time period of HPRINT elapses. HPRINT will be one twelfth of a year. The procedure for the software is outlined in fourteen steps.

1 Reinitialize the random number generator, e.g.:

CALL RANSET (INT(CLOCK(V))).

In other words, we choose the integer that begins the random series according to the time of day.

2 Read in the parameters and the initial values of the variables.
3 Set (a) time $T = 0$, (b) the 'Print time' $TP = 0$, and (c) the 'Stochastic b time' $Tb = 0$. Now we begin the waiting time loop yielding the time evolution of the complete hybrid system.
4 Calculate $\{A_1, A_2, \ldots A_{NE}\}$ the birth and death event rates. They will be functions of the parameters and the present value of the variables, $\vec{X}$.
5 Calculate the sum of the rates:

$$\sum_{E=1}^{NE} A_E$$

6 Calculate WT, the waiting time. (a) Choose a random number $R \in (0, 1)$,

(b): $WT = -\ln(1-R) / (\sum_{E=1}^{NE} A_E)$

7 If $T + WT < TP$, then go to (10). If $T + WT > TP$, then:
 (a) print the values of the variables,
 (b) set $TP = TP + HPRINT$,
 (c) go to (8).
8 If $T + WT > Tb$, then go to (10). If $T + WT > Tb$, then we select the new value of b from a Gaussian whose mean is b_o and whose standard deviation is σ_b.
 (a) Choose two random numbers, R_1 and $R_2 \in (0, 1)$.
 (b) Calculate $G = -2 \ln R_1 * \cos(2 R_2)$.
 (c) $b = b_0 + \sigma b * G$.
 (d) $TB = TB + 1$ year.
 (e) Go to (9).
9 If $T + WT > TEND$, stop.

The order of 7, 8 and 9 is chosen assuming HPRINT $<<1$ year $>>$ TEND. If $T + WT < TP$, we assume that $T + WT < Tb$ and that $T + WT < TEND$, which saves calculating more than one IF/THEN statement with each pass through the waiting time loop.

10 Integrate the variables that undergo continuous change, using the time step $DT = WT$. If WT is small, Euler's method is sufficient, so:

$X_1 \rightarrow X_1 + f_1 * WT$
$X_2 \rightarrow X_2 + f_2 * WT$
$\ldots$

$$X_{Nv} \to X_{Nv} + f_{Nv} \cdot WT$$

If WT is not small, any other initial value integration method such as Runge–Kutta can be used to calculate the continuous evolution.

11 The time is now advanced by letting $T \to T + WT$

12 Select the birth and death event that occurs in this waiting time loop, using the probabilities for each of the possible events:

$$P(E_i) = A_i \ (\sum_{E=1}^{NE} A_E)$$

Specifically,

(a) choose a random number $R_{ES} \in (0, 1)$;

(b) Event E_1 is chosen if $R_{ES} < P(E_1)$. For all other events, $i > 1$, E_k is chosen if:

$$\sum_{i=1}^{k-1} P(E_i) \ < R_{ES} \le \sum_{i=1}^{k} P(E_i)$$

13 Change the variables as the effect of the chosen event. If E_k is chosen:

$$\vec{X}_k \to \vec{X}_k + \Delta \vec{X}_k .$$

14 Go to (4).

In deciding how to order the various steps associated with continuous and discrete processes, we assume that at $T = 0$, the continuous processes begin changing immediately, and then after waiting a time WT, the first birth and death event, occurs. This method may be applied quite generally to any dynamic system involving birth and death processes and continuous change with or without time-dependent or stochastically varying parameters.

NOTE

1   This has been popularized as the 'tragedy of the commons' (Hardin 1968). However, this is incorrect, because community interests generally override Hardin's three assumptions, i.e. that users must pursue personal gain against best interests; the environment is limited such that exploitation exceeds replenishment; the resource is owned by society but open to any user (Stillman 1985).

REFERENCES

Abramowitz, M., and Stegun, I. (1970) *Handbook of Mathematical Functions*, Washington, D.C.: U.S. Department of Commerce, National Bureau of Standards.

Acheson, J. (1981) 'Anthropology of fishing', *Annual Review of Anthropology* 10: 275–316.

Allen, P.M., and McGlade, J.M. (1986) 'Dynamics of discovery and exploitation: the case for the Scotian Shelf groundfish fisheries', *Canadian Journal of Fisheries and Aquatic Sciences* 43: 1187–200.

Allen, P.M., and McGlade, J.M. (1987a) 'Modelling complex human systems: a fisheries example', *European Journal of Operations Research* 30 (2): 147–67.

Allen, P.M., and McGlade, J.M. (1987b) 'Evolutionary drive: the effect of microscopic diversity, error making and noise', *Foundations of Physics* 17 (7): 723–38.

Anderson, K.P., and Ursin, E. (1977) 'A multispecies extension to the Beverton and Holt theory of fishing, with accounts of phosphorus circulation and primary production', *Meddelelser fra Kommissionen for Danmarks Fiskerei-og-Havundersogelser Series Fiskeri* 7: 319–35.

Anderson, L.G. (1986). *The Economics of Fisheries Management*, Baltimore: Johns Hopkins University Press.

Arntz, W.E. and Robles A.P. (1980) *Estudio del potential pesquero de interes para el complejo pesquero 'La Puntilla' (Piso, Peru), Callao, Peru: Programa Cooperative Peruano-Aleman de Investigaciones Pesqueras.*

Bailey, C. (1985) 'The blue revolution: the impact of technological innovations on Third World fisheries', *Rural Sociologist* 5 (4): 259–66.

Bergh, M. (1986) 'The Value of Catch Statistics and Records of Guano Harvests for Managing certain South African Fisheries', doctoral dissertation, University of Cape Town, South Africa.

Berkes, F. (1985) 'Fishermen and "The Tragedy of the Commons"', *Environmental Conservation* 12: 199–206.

Berkes, F. (1987) 'The common property resource problem and the fisheries of Barbados and Jamaica', *Environmental Management* 11 (2): 225–35.

Beverton, R.J.H., and Holt, S.J. (1957) 'On the dynamics of exploited fish populations', *Fisheries Investigations, Ministry of Agriculture, Fisheries and Food, Great Britain*, 2, *Sea Fisheries*, 19: 1–533.

Caddy, J.F., and Gulland, J.A. (1983) 'Historical patterns of fish stocks', *Marine Policy* 7 (4): 267–78.

Caddy, J.F., and Sharp, G.D. (1986) *An Ecological Framework for Marine Fishery Investigations*, Fisheries Technical Paper 283, Rome: Food and Agriculture Organization of the United Nations.

Clark, C.W. (1976) *Mathematical Bioeconomics: the Optimal Management of Renewable Resources*, New York: Wiley.

Clark, C.W. (1985) *Bioeconomic Modelling and Fisheries Management*, New York: Wiley.

Clark, S.H., Overholtz, W.J., and Hennemuth, R.C. (1982) 'Review and assessment of the Georges Bank and Gulf of Maine haddock fishery', *Journal of Northwest Atlantic Fisheries Science* 3: 1–27.

Cox, S.J.B., and Wiebe, P.H. (1979) 'Origins of oceanic plankton in the middle Atlantic bight', *East Coast Marine Science* 9: 509–27.

Daan, N. (1978) 'A review of replacement of depleted stocks by other species and the mechanisms underlying such replacement', *Rapport et procès-verbaux des réunions, Conseil Permanent International pour l'Exploration de la Mer* 177: 405–21.

Duer Stevenson, B. (1986) 'What fishermen think about fishery management', in J.G. Sutinen and L.C. Hanson (eds) *Proceedings of the 10th Annual*

Conference, Center for Oceanic Management Studies, Kingston: University of Rhode Island, pp. 18–20.

Fleminger, A., and Hulseman, K. (1977) 'Geographical range and taxonomic divergence in North Atlantic *Calanus* (*C. Helgolandicus, C. finmarchicus* and *C. glacialis*)', *Marine Biology* 40: 233–48.

Garrett, C. (1972) 'Tidal resonance in the Bay of Fundy and Gulf of Maine', *Nature* 238: 441–3.

Garrett, C.J.R., Keeley, J.R., and Greenberg, D.A. (1978) 'Tidal mixing versus thermal stratification in the Bay of Fundy and Gulf of Maine', *Atmosphere–Ocean* 16: 403–23.

Gatewood, J.B. (1984) 'Cooperation, competition, and synergy: information sharing groups among southeast Alaskan salmon seiners', *American Ethnologist* 11: 350–70.

Gillespie, D.T. (1977) 'Exact stochastic simulation of coupled chemical reactions', *Journal of Physical Chemistry* 81: 2340–61.

Glantz, M.H. (1986) 'Man, state, and fisheries: an inquiry into some social constraints that affect fisheries management', *Ocean Development and International Law* 17: 191–346.

Gordon, H.S. (1954) 'The economic theory of a common property resource: the fishery', *Journal of Political Economy* 62: 124–42.

Gordon, H.R., Clarke, D.K., Mullen, J.L., and Hovis, W.A. (1980) 'Phytoplankton pigments from Nimbus-7 coastal zone color scanner: comparisons with surface measurements', *Science* 210: 63–6.

Halliday, R.G., McGlade, J., Mohn, R., O'Boyle, R.N., and Sinclair, M. (1986) 'Resource and fishery distributions in the Gulf of Maine area in relation to the Subarea 4/5 boundary', *Northwest Atlantic Fishery Organization Scientific Council Studies* 10: 67–92.

Hardin, G. (1968) 'The tragedy of the commons', *Science* 16: 1243–8.

Haydon, P.B., and Dolan, R. (1976) 'Coastal marine fauna and marine climates of the Americas', *Journal of Biogeography*, 3: 71–81.

Horsthemke, W., and Lefever, R. (1984) *Noise Induced Transitions: Theory and Applications in Physics, Chemistry and Biology*, Springer Series in Synergetics 15, New York: Springer.

Huberman, B.A., and Hogg, T. (1988) 'The behavior of computational ecologies', in B. Huberman (ed.) *The Ecology of Computation*, Amsterdam: Elsevier.

Iles, T.D. (1986) 'Interaction of external and internal factors in relation to recruitment', Northwest Atlantic Fisheries Organization scientific document 86/119.

Koeller, P.A., Hurley, P., Perley, P., and Neilson, V.D. (1986) 'Juvenile fish surveys on the Scotian Shelf: implications for year–class size assessments, *Rapport et procès-verbaux des rèunions, Conseil Permanent International pour l'Exploration de la Mer* 43: 59–76.

Krasner, S.D. (1982) 'Structural causes and regime consequences: regimes as intervening variables', *International Organization* 36: 185–205.

Laevastu, T., and Larkins, H.A. (1981) *Marine Fisheries Ecosystem: its Quantitative Evaluation and Management*. Farnham: Fishing News Books.

Leap, W.L. (1977) 'Maritime subsistence in anthropological perspective: a statement of priorities', in M.E. Smith (ed.) *Those who Live from the Sea*, St Paul: West Publishing, pp. 251–63.

MacCall, A. (1986) 'Rethinking research for fishery and ecosystem management', in J.G. Sutinen and A.C. Hanson (eds) *Proceedings of the 10th*

Annual Conference, Center for Oceanic Management Studies, Kingston: University of Rhode Island, pp. 179–93.

McCay, B. (1978) 'Systems ecology, people ecology, and the anthropology of fishing communities', *Human Ecology* 6: 397–422.

McGlade, J.M. (1987) 'The influence of Gulf Stream gyres on recruitment in pollock (*Pollachius virens*)', in R.I. Perry and K.T. Frank (eds) *Environmental Effects on Recruitment to Canadian Atlantic Fish Stocks*, Canadian Technical Report of Fisheries and Aquatic Sciences.

McGlade, J.M., and Allen, P.M. (1985) 'The fishing industry as a complex system', in R. Mahon (ed.) *Toward the Inclusion of Fishery Interactions in Management Advice*, Canadian Technical Report of Fisheries and Aquatic Sciences 1347, pp. 209–16.

Mahon, R., and Neilson, J.D. (1987) 'Diet changes in Scotian Shelf haddock during the pelagic and demersal phases of the first year of life', *Marine Ecology Progress* series 37: 123–30.

Mahon, R., Smith, R.W., Bernstein, B.B. and Scott, J.S. (1984) *Spatial and Temporal Patterns of Groundfish Distribution on the Scotian Shelf and in the Bay of Fundy, 1970–1981*, Canadian Technical Report of Fisheries and Aquatic Sciences 1300.

Majone, G. (1986) 'International institutions and the environment', in W.C. Clark and R.E. Munn (eds) *Sustainable Development of the Biosphere*, Cambridge: Cambridge University Press, pp. 351–9.

May, R.M. (1974) *Stability and Complexity in Model Ecosystems*, Princeton: University Press, New Jersey.

May, R.M. (ed.) (1984) *Exploitation of Marine Communities*, Dahlem Workshop Reports, Life Sciences Research Reports 32, Berlin: Springer.

Maynard Smith, J. (1974) *Models in Ecology*, Cambridge: Cambridge University Press.

Mills, E.L. (1980) 'The structure and dynamics of shelf and slope ecosystems off the north-east coast of North America', in K.R. Tenore and B.C. Coull (eds) *Marine Benthic Dynamics*, Columbia: University of South Carolina Press, pp. 25–47.

Nisbet, R.M. and Gurney, W.S.C. (1982) *Modelling Fluctuating Populations*, Chichester: Wiley.

O'Boyle, R., McMillan, J. and White, G. (1984) 'The 4X haddock resource: a problem in supply and demand', Canadian Atlantic Fisheries Scientific Advisory Committee research document 84/100.

O'Boyle, R., and Wallace, D. (1986) 'An evaluation of the population dynamics of 4X haddock during 1962–85 with yield projected to 1987', Canadian Atlantic Fisheries Scientific Advisory Committee research document 86/98.

Orbach, M.L. (1977) *Hunters, Seamen and Entrepreneurs: the Tuna Fishermen of San Diego*, Berkeley: University of California Press.

O'Reilly, J.E., and Evans-Zetlin, C. (1982) 'A comparison of the abundance (Chlorophyll a) and size composition of the phytoplankton communities in twenty Subareas of Georges Bank and surrounding water', International Council for the Exploration of the Sea CM 1982/L.

Pauly, D. (1979) 'Theory and management of tropical multispecies stocks: a review with emphasis on the Southeast Asian demersal fisheries', *ICLARM Studies and Reviews 1*, Manila: International Center for Living Aquatic Resources Management.

Peer, D., Wildish, D.J., Wilson, A.J., Hines, J., and Dadswell, M. (1980) *Sublittoral Macro-fauna of the Lower Bay of Fundy*, Canadian Technical Report of Fisheries and Aquatic Sciences 981.

Pielou, E.C. (1969) *An Introduction to Mathematical Ecology*, New York: Wiley.

Puchala, D.J., and Hopkins, R.F. (1982) 'International regimens: lessons from inductive analysis', *International Organization* 36: 245–76.

Ricklefs, R.E. (1987) 'Community diversity: relative roles of local and regional processes', *Science* 235: 167–71.

Schaefer, M.B. (1954) 'Some aspects of the dynamics of populations important to the management of the commercial marine fisheries', *Bulletin of the Inter-American Tropical Tuna Commission* 1 (2): 27–56.

Schaefer, M.B. (1957) 'A study of the dynamics of the fishery for yellowfin tuna in the eastern tropical Pacific Ocean', *Bulletin of the Inter-American Tropical Tuna Commission* 2: 247–352.

Sherman, K. (1978) 'Ecological implications of biomass changes in the northwest Atlantic', International Council for the Exploration of the Sea CM 1978/L.

Sherman, K. (1986) 'Measurement strategies for monitoring and forecasting variability in large marine ecosystems', in K. Sherman and L.M. Alexander (eds) *Variability and Management of Large Marine Ecosystems*, American Association for the Advancement of Science Selected Symposium 99, Boulder: Westview Press, 203–36.

Sinclair, P. (1983) 'Fishermen divided: the impact of limit entry licensing in northwest Newfoundland', *Human Organization* 42: 307–31.

Sissenwine, M.P. (1984) 'The uncertain environment of fishery scientists and managers', *Marine Resource Economics* 1: 1–30.

Smith, V.L. (1969) 'On models of commercial fishing', *Journal of Political Economics* 77: 81–98.

Stillman, P.G. (1985) 'The tragedy of the commons: a reanalysis', *Alternatives* 4 (2): 12–15.

Topliss, B.J. (1982) 'Water color in eastern Canadian inshore areas', *Northwest Atlantic Fisheries Organization Scientific Council Studies* 4: 63–7.

United National (1982) *World Fisheries and the Law of the Sea*, Rome: Food and Agriculture Organization of the United Nations.

Ursin, E. (1982) 'Stability and variability in the marine ecosystem', *Dana* 2: 51–67.

Walters, C.J. (1986) *Adaptive Management of Renewable Resources*, New York: Macmillan.

Weibe, P.M. (1983) 'Rings of the Gulf Stream', *Scientific American* (March) 60–70.

Wigley, R.L. (1968) 'Benthic invertebrates of the New England fishing banks', *Underwater Naturalist* 5 (1): 1–13.

Young, O.R. (1982) *Resource Regimes*, Berkeley: University of California Press.

ACKNOWLEDGEMENTS

We would like to thank P. Allen, R.N. O'Boyle and members of the Prospero Project of the International Federation of Institutes for Advanced Study for discussions about this work. The work was supported by the Marine Fish Division, Department of Fisheries and Oceans, Canada.

Index